my **revisi⊙n** notes

WJEC GCSE

Food and Nutrition (Wales)

Helen Buckland

HODDER
EDUCATION
AN HACHETTE UK COMPANY

The Publishers would like to thank the following for permission to reproduce copyright material.

Photo credits
p.7 © 1997 C Squared Studios/Photodisc/Getty Images/ Eat, Drink, Dine 48; **p. 9** © Imagestate Media (John Foxx)/Animals, Nature & Scenics Vol 30; **p.10** © Ingram Publishing Limited/Ingram Image Library 500-Food; **p.11** © Photodisc/Getty Images/World Commerce & Travel 5; **p.12** © Kevin Britland/Alamy Stock Photo; **p.13** © yellowj – Fotolia; **p.15** © HABAKUKKOLO - iStockphoto via Thinkstock/Getty Images; **p.16** © FOOD-pictures – Fotolia; **p.17** © Clynt Garnham Food & Drink/Alamy Stock Photo; **p.18** © BrandX/Getty Images/Food and Textures CD X025; **p.20** © WavebreakmediaMicro – Fotolia; **p.21** © giovanni1232 - iStockphoto via Thinkstock/Getty Images; **p.22** © Photolibrary.Com; © Vikram Raghuvanshi/iStockphoto.com; **p.24** © Dušan Zidar – Fotolia; © Monkey Business – Fotolia; **p.29** © matka_Wariatka – Fotolia; **p.32** © Fuse - iStockphoto via Thinkstock/Getty Images; **p.36** © Okea – Fotolia; p.37 © Maridav – Fotolia; **p.39** © alinamd – Fotolia; **p.41** © Monkey Business – Fotolia; **p.43** © volff – Fotolia; **p.44** © Michael Neelon(misc)/Alamy Stock Photo; **p.47** © mezzotint_fotolia – Fotolia; **p.50** © Ross Land - Getty Images; **p.53** © BlueOrange Studio – Fotolia; **p.54** © MELBA PHOTO AGENCY/Alamy/Ingredients CD0163D; **p.55** © Awe Inspiring Images – Fotolia; © Paul_Brighton - iStockphoto via Thinkstock/ Getty Images; © Tatiana Volgutova - iStockphoto via Thinkstock/Getty Images; © Billy_Fam - iStockphoto via Thinkstock/Getty Images; **p.56** © KucherAV - iStockphoto via Thinkstock/Getty Images; **p.58** © LOU63 - iStockphoto via Thinkstock/Getty Images; **p.59** © Hodder Education; **p.62** © marilyn barbone – Fotolia; **p.67** © Zoltan Fabian –Shutterstock; © Antonsov85 – Shutterstock; **p.68** © Kondor83 - iStockphoto via Thinkstock/Getty Images; **p.69** © F1online digitale Bildagentur GmbH/Alamy Stock Photo; **p.70** © Cultura RM/Alamy Stock Photo; **p.72** © geoffbooth19 – Fotolia; **p.73** © Gannet77 - iStockphoto via Thinkstock/Getty Images; © Stockbyte/ Photolibrary Group Ltd/ Environmental Issues DV 48; **p.75** © Getty Images/Image Source - OurThreatened Environment IS236; **p.81** © Joe Gough – Fotolia; **p.82** © Mat Hayward – Shutterstock; © M.studio – Fotolia; **p.85** © Ryan McVay - iStockphoto via Thinkstock/Getty Images; **p.93** © funkyfood London - Paul Williams/ Alamy Stock Photo; **p.94** © MediablitzImages – Fotolia; **p.95** © Monkey Business – Fotolia; **p.101** © Ingram Publishing - iStockphoto via Thinkstock/Getty Images; © Thomas Northcut - iStockphoto via Thinkstock/Getty Images; **p.102** © Oleg Pchelov – Shutterstock; © Ciaran Walsh/ iStockphoto.com; **p.103** © Andrew Callaghan/ Hodder Education; **p.104** © Ian O'Leary/Getty Images; **p.105** © Serghei Starus - iStockphoto via Thinkstock/Getty Images; © Penny Burt - iStockphoto via Thinkstock/Getty Images; © Lee lian Chong - iStockphoto via Thinkstock/Getty Images; © Uncle_ Bob - iStockphoto via Thinkstock/Getty Images; © zhaubasar - iStockphoto via Thinkstock/Getty Images; **p.107** © Doug Steley A/Alamy Stock Photo; **p.109** © Alex Segre/Alamy Stock Photo; **p.110** © ranplett/iStockphoto; **p.111** © HLPhoto – Fotolia; **p.113** © moreimages - Shutterstock

Acknowledgements
Every effort has been made to trace all copyright holders, but if any have been inadvertently overlooked, the Publishers will be pleased to make the necessary arrangements at the first opportunity.

Although every effort has been made to ensure that website addresses are correct at time of going to press, Hodder Education cannot be held responsible for the content of any website mentioned in this book. It is sometimes possible to find a relocated web page by typing in the address of the home page for a website in the URL window of your browser.

Hachette UK's policy is to use papers that are natural, renewable and recyclable products and made from wood grown in sustainable forests. The logging and manufacturing processes are expected to conform to the environmental regulations of the country of origin.

Orders: please contact Bookpoint Ltd, 130 Park Drive, Milton Park, Abingdon, Oxon OX14 4SE. Telephone: (44) 01235 827720. Fax: (44) 01235 400454. Email education@bookpoint.co.uk Lines are open from 9 a.m. to 5 p.m., Monday to Saturday, with a 24-hour message answering service. You can also order through our website: www.hoddereducation.co.uk

ISBN: 978 1 4718 8540 2

Get the most from this book

Everyone has to decide their own revision strategy, but it is essential to review your work, learn key facts and test your understanding. These Revision Notes will help you to do that in a planned way, topic by topic. You can check your progress by ticking off each section as you revise.

Tick to track your progress

Use the revision planner on pages 4–6 to plan your revision, topic by topic. Tick each box when you have:

● revised and understood a topic
● tested yourself
● practised Now test yourself questions and gone online to check your answers and complete the quick quizzes.

You can also keep track of your revision by ticking off each topic heading in the book. You may find it helpful to add your own notes as you work through each topic.

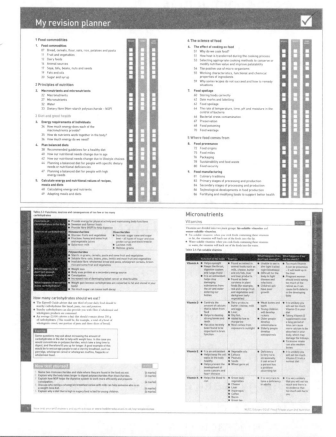

Features to help you succeed

Exam tips

Expert tips are given throughout the book to help you polish your exam technique in order to maximise your chances in the exam.

Typical mistakes

The authors identify the typical mistakes candidates make and explain how you can avoid them.

Now test yourself

These short, knowledge-based questions provide the first step in testing your learning. Answers are at the back of the book.

Key words

Key words from the specification are highlighted in bold throughout the book.

Online

Go online to try out the extra quick quizzes at
www.hoddereducation.co.uk/myrevisionnotes

My revision planner

1 Food commodities

1. Food commodities
07 Bread, cereals, flour, oats, rice, potatoes and pasta
11 Fruit and vegetables
12 Dairy foods
14 Animal sources
17 Soya, tofu, beans, nuts and seeds
19 Fats and oils
20 Sugar and syrup

2 Principles of nutrition

2. Macronutrients and micronutrients
22 Macronutrients
27 Micronutrients
32 Water
33 Dietary fibre (Non-starch polysaccharide – NSP)

3 Diet and good health

3. Energy requirements of individuals
34 How much energy does each of the macronutrients provide?
35 How do nutrients work together in the body?
36 How much energy do we need?

4. Plan balanced diets
38 Recommended guidelines for a healthy diet
40 How our nutritional needs change due to age
42 How our nutritional needs change due to lifestyle choices
44 Planning a balanced diet for people with specific dietary needs or nutritional deficiencies
47 Planning a balanced diet for people with high energy needs

5. Calculate energy and nutritional values of recipes, meals and diets
48 Calculating energy and nutrients
49 Adapting meals and diets

REVISED   TESTED   EXAM READY

Now test yourself answers and quick quizzes at **www.hoddereducation.co.uk/myrevisionnotes**

4 The science of food

6. **The effect of cooking on food**

- 51 Why do we cook food?
- 51 How heat is transferred during the cooking process
- 53 Selecting appropriate cooking methods to conserve or modify nutritive value and improve palatability
- 54 The positive use of micro-organisms
- 55 Working characteristics, functional and chemical properties of ingredients
- 59 Why some recipes do not succeed and how to remedy situations

7. **Food spoilage**

- 60 Storing foods correctly
- 62 Date marks and labelling
- 62 Food spoilage
- 64 The role of temperature, time, pH and moisture in the control of bacteria
- 66 Bacterial cross-contamination
- 67 Preservation
- 68 Food poisoning
- 70 Food wastage

5 Where food comes from

8. **Food provenance**

- 72 Food origins
- 75 Food miles
- 76 Packaging
- 78 Sustainability and food waste
- 80 Food security

9. **Food manufacturing**

- 81 Culinary traditions
- 83 Primary stages of processing and production
- 84 Secondary stages of processing and production
- 86 Technological developments in food production
- 86 Fortifying and modifying foods to support better health

REVISED   TESTED   EXAM READY

6 Cooking and food preparation

10. Factors affecting food choice

		REVISED	TESTED	EXAM READY
89	Sensory perception	☐	☐	☐
91	Tasting panels and preference testing	☐	☐	☐
92	Factors that affect food choices	☐	☐	☐
95	The choices that people make about foods, based on religion, culture or ethical belief, medical reasons or personal choices	☐	☐	☐
95	How to make informed choices about food and drink to achieve a varied and balanced diet	☐	☐	☐
98	Food labelling	☐	☐	☐

11. Preparation and cooking techniques

100	Planning for cooking a single dish or a number of dishes	☐	☐	☐
101	Preparation of ingredients	☐	☐	☐
102	Cooking a selection of recipes	☐	☐	☐
105	Presenting a selection of recipes	☐	☐	☐
106	Working safely	☐	☐	☐
108	Using sensory descriptors	☐	☐	☐

12. Developing recipes and meals

109	The influence of lifestyle and consumer choice when adapting or developing meals and recipes	☐	☐	☐
110	Adaptations to recipes to address current dietary advice	☐	☐	☐
111	Considering nutritional needs and food choices when selecting recipes	☐	☐	☐
112	Reviewing and making improvements to recipes	☐	☐	☐
112	Manage the time and cost of recipes	☐	☐	☐
112	Using testing and sensory evaluation skills	☐	☐	☐
113	Explaining, justifying and presenting ideas about chosen recipes and cooking methods	☐	☐	☐
114	Making decisions about which techniques are appropriate to use during preparation and cooking	☐	☐	☐

Success in the examination

Sample examination questions, model answers and mark schemes

Now test yourself answers at www.hoddereducation.co.uk/myrevisionnotes

1 Food Commodities

Food commodities are the basic foods that make up our daily diet.

Bread, cereals, flour, oats, rice, potatoes and pasta

Bread

REVISED

- Bread is eaten regularly in our diet and contributes in a major way, so it is a **staple food**.
- It can be used in many ways.
- It can be sweet or savoury or flavoured, for example, with cheese, or have extra toppings added, for example, poppy seeds.

Figure 1.1 Bread products

Bread is made from dough. The ingredients are:
- flour
- salt
- yeast
- liquid.

(The liquid is usually water, but can be milk.)

The ingredients are mixed together and kneaded, left to rise, kneaded again, shaped, **proved** (left to rise) and baked.

Nutritional value of bread

Bread is in the starchy carbohydrate section of the Eatwell Guide. It also contains some protein, B-group vitamins, calcium and iron. Wholemeal bread contains fibre.

Storage of bread

- Fresh bread: in a bread bin or sealed paper bag.
- Sliced supermarket bread: in a plastic bag.
- Bread can be frozen for up to two months.
- Bread goes dry and stale quite quickly.

Cereals

REVISED

In the UK we grow wheat, barley, oats and rye. Most of these are processed into other foods before we eat them.

Figure 1.2 Different cereal plants

Nutritional value of cereals

Cereals contain fibre, carbohydrates, Low Biological Value (LBV) protein (plant protein that does not contain all of the essential amino acids), B-group vitamins, Vitamin E, fat and iron.

Storage of cereals

Cereals can become stale, contaminated with bacteria and moulds, or develop different odours if not stored correctly.
- Store cereals in a cool, dry place in an airtight container.
- Keep old and new cereals separate.
- Always check the sell-by or use-by date.

Flour

REVISED

This is made from wheat or rye. There are two main types:
- Strong flour is used for bread making. This is made from hard, winter wheat.
- Weak flour is used for cakes, pastry and biscuits. This is made from spring wheat.

Nutritional value of flour

- Strong flour contains **gluten**, which is a protein. This will be stretched when bread dough is kneaded, and forms the structure of the bread.
- White flour in the UK is **fortified** with iron, calcium and the B-group vitamins Thiamine and Niacin. These are all lost during the processing of wheat into flour.
- Wholemeal flour has all of the bran from the wheat, so has 100% **extraction rate**.
- White flour has the bran, germ, fat and some of the minerals removed, so has a 75% extraction rate.

Storage of flour

- Store flour in a cool, dry place in the original packaging in a sealed container to prevent weevils infesting the flour.
- Check the use-by date before use. Wholemeal flour still contains fat so can become rancid with age.
- Never mix old and new flour.

Oats

REVISED

Before oats can be used, a protective husk around the grain has to be removed.

- Oats can be crushed or rolled to form oatmeal, which is used for porridge or to make flapjacks.
- Jumbo oat flakes can be used for breakfast cereals such as muesli.
- Fine oat flour can be made by grinding the grains, or by processing them in a food processor. The flour can be used for baked foods such as biscuits and scones.

Nutritional value of oats

Oats contain carbohydrates, plus smaller amounts of protein, fat, calcium, iron and some B-group vitamins. Starchy carbohydrates from oats provide a slow releasing energy source.

Storage of oats

- In a dry, cool place
- In an airtight container after opening.

Figure 1.3 Oats

Rice

REVISED

- Rice is a staple food, as it is eaten all over the world as a main part of the diet.
- It has an outer husk removed during processing.
- It can be used for sweet and savoury foods.
- It can be boiled, baked or stir-fried.
- It can be bought as short-grain or long-grain varieties.

Nutritional value of rice

Rice is a carbohydrate food, so is an excellent energy source.

Storage of rice

Store in a dry, cool place, in an airtight container after opening.

Potatoes

- Potatoes are a staple food.
- They are grown in the UK and come in different varieties, which can be used for different methods of cooking.
- Potatoes can be baked, boiled, roasted or fried.

Nutritional value of potatoes

Potatoes contain starchy carbohydrates, Vitamin C, Vitamin B_6 and Thiamine, and the skin contains fibre.

Storage of potatoes

- Potatoes should be stored in a cool, dark, dry, airy place. Exposure to light can cause them to turn green, which makes them toxic. The green parts should be removed before cooking.
- Storing potatoes in plastic bags makes them sweat and turn mouldy.

Figure 1.4 Different types of potatoes

Pasta

- Pasta is made from a strong type of wheat that is called **durum wheat**. This wheat contains more protein.
- Pasta is made with durum wheat flour, water, salt and sometimes eggs and oil.
- Pasta can also be coloured with spinach, tomato and squid ink.
- Pasta can be bought as dried or fresh in many different shapes.

Nutritional value of pasta

Pasta contains starchy carbohydrates. Wholemeal types contain fibre.

Storage of pasta

- Dried pasta: keep in an airtight container once opened.
- Fresh pasta: keep in the fridge.
- Homemade pasta: dry and store in the fridge in an airtight container.
- Fresh and homemade pasta can be frozen.

> **Exam tip**
>
> A question on bread, cereals, flour, oats, rice, potatoes and pasta may ask about staple foods in different countries. Asian countries have rice as a staple food because the climate is right for growing rice. The UK has the correct climate for growing wheat and potatoes, so those form a large part of our diet. You will need to think of reasons why a particular staple food is eaten.

Now test yourself

1 Discuss the nutritional benefits of including potatoes in the diet. [4 marks]
2 Explain why white flour is fortified in the UK. [2 marks]
3 List two ways that rice can be cooked, and suggest a recipe for each of your chosen ways. [4 marks]
4 Give one reason why eating more wholemeal pasta may be beneficial in someone's diet. [2 marks]

Fruit and vegetables

- Fruits and vegetables can be bought as fresh, frozen, canned, juiced and dried.
- Some are **seasonal** and only available at certain times of the year.
- Some fruits and vegetables are grown in the UK.
- Many fruits and vegetables are imported from abroad.

Fruit

Nutritional value of fruit

Eating a variety of different types and colours of fruit will provide vitamins A, C and E, carbohydrates, fibre and some minerals.

Storage of fruit

- Citrus fruits such as oranges, lemons and limes: in a cool, dry place.
- Berry fruits such as raspberries and strawberries: in the fridge.
- Hard fruits such as apples and pears: keep out of direct sunlight or in the fridge.
- Stone fruits such as plums and peaches: in the fruit bowl or the fridge.
- Exotic fruits such as bananas and pineapples: in the fruit bowl.

Figure 1.5 Tropical fruits

Vegetables

Types of vegetables

Table 1.1 Types of vegetable

Type of vegetable	Examples
Leafy	Cabbage, lettuce
Tubers	Potatoes
Root	Carrots, turnips
Stems	Asparagus, celery
Flowers	Cauliflower, broccoli
Fruits and seeds	Peas, courgettes
Fungi	Mushrooms

- Wales is well known for growing leeks, which are a staple vegetable in Wales and form the basis of a dish called Glamorgan Sausages.
- In Wales, an edible seaweed known as laver is gathered and processed commercially. It is used to produce bara lawr or laverbread, which is usually eaten sprinkled with oatmeal, then warmed in hot bacon fat and served with bacon for breakfast or supper. The seaweed itself can be found in some parts of the west coast, clinging to the rocks at low tide.

Nutritional value of vegetables

Eating a variety of different types and colours of vegetables will provide vitamins A, C and E, carbohydrates, fibre and some minerals. Frozen vegetables can be as nutritious as fresh vegetables.

Storage of vegetables

- Most vegetables should be stored in cool, dry, well-ventilated places.
- Salad and green vegetables should be stored in the fridge. Green leafy vegetables will quickly lose vitamin C as they age.
- Vegetables should be eaten as fresh as possible to get maximum nutritional value.

Dairy foods

Milk

REVISED

- The most common form of milk consumed in the UK is cow's milk. This comes in several varieties such as whole milk, semi-skimmed and skimmed milk.
- Other forms of milk are available, such as goat's milk, or alternative non-dairy forms such as rice, almond and soya milk.
- Milk is treated to make it safe to drink by killing harmful bacteria, using heat treatments.
- Pasteurised milk is heated to 72°C for 15 seconds then rapidly cooled to below 10°C and put into cartons or bottles.
- UHT milk, or long-life milk, is heated to 132°C for 1 minute, cooled rapidly and packed in sterile conditions. This lasts for many months, until opened.

Figure 1.6 Milk

Nutritional value of milk

Milk contains HBV (High Biological Value) protein, fat, sugar in the form of lactose, vitamins A, D, some of the B-group vitamins and a little Vitamin C, calcium, potassium and a small amount of iron.

Storage of milk

- Fresh milk: in the fridge and consumed by its use-by date.
- UHT cartons: in a cool, dry place; once opened treat as fresh milk and store in the fridge.

Cheese

- Cheese is made from fermented milk.
- Enzymes are added to **denature** the protein, and produce a solid. Then flavourings are added to produce different types of cheese.
- There are hundreds of types of cheeses produced both in the UK and around the world.
- Cheese can provide flavour, colour and texture to a dish, as well as valuable nutrients.
- Cheese has long been a traditional food of Wales and award-winning varieties, from the more famous Caerphilly, Tintern, and Y Fenni to the likes of Black Bomber and Perl Las, are produced using Welsh milk.

Figure 1.7 Different types of cheese

Nutritional value of cheese

Cheese contains HBV protein, calcium, varying amounts of potassium and sodium (depending on the type of cheese), vitamins A, D and some B-group vitamins (depending on the type of cheese). It has a high fat content.

Storage of cheese

- Cheese must be stored in the fridge.
- Hard-pressed cheeses last a long time, but must be wrapped, as they will dry out.
- Soft cheeses have a shorter shelf life and should be consumed within a few days.

Yoghurt

- Yoghurt is made by adding 'friendly' bacteria to milk.
- This causes it to ferment by changing the sugar in the milk (lactose) to lactic acid, which denatures the protein and causes it to set.

Types of yoghurt

- Yoghurt can be made from different types of milk.
- Set yoghurt is set in the pot in which it is sold. Flavourings, fruit and sugar are often added.
- Live yoghurt contains live bacteria, which can be beneficial to the working of your digestive system.
- Greek yoghurt is thicker, has a higher fat content and is made from cow's or sheep's milk.

Nutritional value of yoghurt

Yoghurt contains HBV protein, varied amounts of fat (depending on the type of yoghurt), calcium, sugar (lactose), vitamins A and D, some of the B-group vitamins and Vitamin E if it is whole-milk yoghurt.

Storage of yoghurt

- Store in the fridge.
- Eat by its use-by date.

> **Exam tip**
>
> You could be asked to suggest how you would substitute dairy products with alternatives to provide an ingredient that is suitable for someone who is lactose intolerant, or is a vegan.
>
> Remember to explain in detail why the person cannot eat the ingredient in the original recipe, what you are using as a substitute, and why the replacement ingredient is suitable for the person.

Now test yourself

1 List three nutrients found in milk. [3 marks]
2 Suggest one savoury and one sweet recipe where milk is one of the main ingredients. [2 marks]
3 Explain how yoghurt is made. [4 marks]
4 Name two Welsh cheeses. [2 marks]

Animal sources

Meat

- Animals used for meat in the UK are cows (for beef or veal), sheep (for lamb or mutton) and pigs (for pork, bacon, gammon or ham).
- Meat is made of muscle fibres, connective tissue and fat.
- The fat is either **visible**, for example, the fat that can be seen around the edge of a steak, or **invisible**, which is found in the connective tissue and is known as marbling.
- The length of muscle fibre determines how tough the meat is: longer fibres in the legs will be tougher than those in the back. Tougher cuts of meat will need long, slow cooking to make it tender.
- Meat is a high-risk food that must be prepared and cooked correctly to avoid food poisoning.
- Welsh lamb and beef have Protected Geographical Indication (PGI) from the European Commission. Lambs and calves born in Wales are tagged and logged from birth so can be identified at every stage of their lives and during production of meat products after being slaughtered.
- Gower Salt Marsh Lamb is grazed on the salt marshes.
- Welsh Black beef is also a recognised individual meat from Wales.

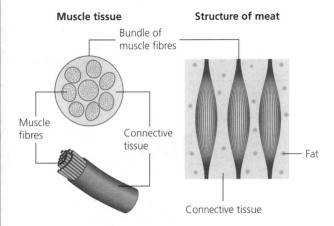

Figure 1.8 Structure of meat

Nutritional value of meat

Meat contains HBV protein, varying amounts of fat, vitamins A and D, some of the B-group vitamins (it is a good source of Vitamin B_{12}) and iron.

Storage of meat

- Raw meat: in the fridge on the bottom shelf. It should be in a covered container and used by its use-by date. If frozen, it should be well wrapped and defrosted thoroughly before cooking and should never be refrozen.
- Cooked meat: be cool within 1.5 hours, cover and put in the fridge. Store above raw meat in the fridge to avoid cross-contamination.

Fish

There are three main types of fish:
- **White fish**, for example, cod, haddock and plaice.
- **Oily fish**, for example, mackerel, sardines and fresh tuna.
- **Shellfish**, for example, crabs and lobster.
- Wales is famous for shellfish called cockles, which are gathered on the coast by small, local businesses, particularly on the Gower coast, near Penclawdd. Laws protect these cockles and licences are issued for their gathering, with only a certain number of licences issued per year. The gathering must still be done by hand with a rake and riddle.

Figure 1.9 Different types of fish

Fish is made of muscle and connective tissue. The muscle fibres are short, so fish is cooked quickly and is tender.
- Fish can be bought fresh, frozen, smoked, canned or dried.
- Fish can be cooked in a number of ways: baked, fried, grilled or poached.

Nutritional value of fish

Fish contains HBV protein, essential fatty acids (in oily fish), a good source of vitamins A and D (in oily fish) and calcium (if the bones are eaten, for example, in canned sardines). Fish is low in fat, but shellfish may contain high levels of cholesterol.

Storage of fish

- Fresh fish: in the fridge. Use as soon as possible after buying, as it will go off quickly.
- Frozen fish: fully defrosted before cooking. Do not refreeze raw fish.

Poultry

- Poultry refers to turkey, chicken, duck, goose, guinea fowl and pigeon.
- Chicken is the most popular poultry in the UK.
- Poultry meat is made of muscle fibres, fat and connective tissue.
- Poultry can be cooked in several ways, for example, roasting, baking, frying, poaching and grilling.
- Poultry is a high-risk food that may contain salmonella bacteria.
- Cook poultry thoroughly until the centre reaches 72°C for at least two minutes.

Figure 1.10 Chicken

Nutritional value of poultry

Poultry contains HBV protein, some fat (but less than meat), vitamins A and D and some of the B-group vitamins.

Storage of poultry

- Raw poultry should be stored in the fridge on the bottom shelf. Keep it covered to prevent cross-contamination.
- Raw poultry can be frozen. Thoroughly defrost frozen poultry before cooking.
- Use poultry before its use-by date.

Eggs

- Eggs can be from hens, ducks, geese and quails in the UK.
- They come in different sizes.
- Eggs can be fried, poached, boiled, baked or scrambled.
- Eggs are used in many recipes to bind, set, enrich and aerate.
- Eggs can be from 'enriched' cage farm hens, where the hens are kept in cages.
- **Free-range eggs** come from hens that are allowed to roam outside.
- **Barn eggs** are from hens that can walk around inside a barn.

Eggs are a high-risk food and must be cooked properly to kill any salmonella bacteria.

Nutritional value of eggs

Table 1.2 Nutritional value of eggs

Egg white	Egg yolk
HBV protein	HBV protein
B-group vitamins	Vitamins A, D and E
	Iron
	Fat

Storage of eggs

- Store in the fridge with the pointed end down.
- Keep away from strong smelling foods.
- Use before the best-before date.

Gelatine

- Gelatine is a flavourless food that is made from an animal source and is derived from collagen boiled down from bones, skins and tendons.
- It is commonly used as a gelling agent in food, for example, in desserts or in fruit jelly sweets such as jelly babies.
- It is used as a stabiliser, thickener or texturiser in foods such as yoghurt, cream cheese and margarine.
- It is used in reduced-fat foods to simulate the mouth-feel of fat and to create volume.
- It is used for the clarification of juices, such as apple juice, and vinegar.
- It comes in leaf and powdered forms.

Exam tip

Questions on meat, fish and eggs may focus on the nutritional content of these commodities. The question could be about which nutrients are contained in oily fish. Remember to differentiate between oily fish and white fish, explaining about the importance of the essential fatty acids found in oily fish.

Now test yourself

1 Explain why government recommendations are to eat at least two portions of fish a week, one of which should be oily fish. [4 marks]
2 State the reason that stewing steak is tougher than fillet steak, and give one way you would cook the stewing steak to make it tender. [3 marks]
3 List the three main types of fish, giving one example of each type. [3 marks]
4 Discuss reasons that someone may prefer to buy free-range hen eggs over caged hen eggs. [6 marks]
5 Explain the difference between Welsh salt marsh lamb and English lamb. [3 marks]
6 Explain why only a certain number of licences are issued each year to allow cockles to be gathered on the Gower coast. [4 marks]

Soya, tofu, beans, nuts and seeds

Soya

REVISED

- It is made from soya bean pods, a member of the legume family.
- Soya is made into various products that can replace animal sources of protein such as tempeh, miso, soya milk and soy sauce.
- Soya beans can be bought in dried, canned and fresh (edamame beans) varieties.

Nutritional value of soya

Soya beans contain HBV protein, calcium, magnesium and are high in fibre.

Storage of soya products

- Fresh edamame beans: in the fridge.
- Tempeh and miso: in the fridge.
- Dried and tinned soya products: in a cool, dry place until cooked or opened, and then store in the fridge.

Tofu

REVISED

- Tofu is also known as bean curd.
- Tofu is made from fresh soya milk that has been curdled and pressed into a block, similar to the way cheese is made.
- Tofu does not have much flavour, so it is used in curries and stir-fries where it is cooked with stronger flavoured foods.
- Tofu can be marinated to absorb flavours before cooking.

Nutritional value of tofu

Tofu contains HBV protein, iron, calcium and some of the B-group vitamins.

Storage of tofu

- It is a chilled product so must be stored in the fridge.
- It can be frozen in its original packet, and fully defrosted before cooking.

Mycoprotein (Quorn) products

REVISED

- Quorn products are not animal protein, but are meat substitute products. All Quorn foods are made from a mycoprotein that comes from a fungus.
- Quorn products are a source of HBV protein that contains all of the essential amino acids needed by the body.
- Quorn products that are made without egg albumin are now available for vegans.

Figure 1.11 Quorn products

Beans

- Beans are commonly known as pulse vegetables.
- They are sold in fresh, frozen, canned and dried forms.
- They include haricot beans (baked beans), lentils and split peas.
- They add flavour, bulk, colour and texture to a dish.

Nutritional value of beans

Beans contain LBV (low biological value) protein, some carbohydrates, iron and calcium, some of the B-group vitamins and fibre.

Storage of beans

- Fresh beans: in the fridge and use by their use-by date.
- Dried beans: in airtight containers in a cool, dry place. Once cooked store in the fridge.
- Frozen beans: in the freezer; defrost before cooking.
- Canned beans: in a cool, dry cupboard until opened, and then store in the fridge.

Figure 1.12 Different types of bean

Nuts

Nuts are available in many forms: whole, chopped, flaked, roasted, salted or blanched. These can be:
- edible kernels such as almonds and walnuts
- from inside a dry shell, such as cashew or chestnuts
- pulses, such as peanuts
- seeds, such as brazil or pine nuts
- used in both savoury and sweet dishes.

Nutritional value of nuts

Nuts contain LBV protein, some of the B-group vitamins and fibre. Some nuts contain high fat content, calcium and iron.

Storage of nuts

- Store nuts in airtight containers, as nuts can become rancid if exposed to air.
- Use nuts by their best-before date.

Seeds

- Seeds can be used as a topping for salads, bread or cakes.
- They can be toasted or roasted to add texture to foods.
- They can be used to manufacture oils for salad dressings or cooking.
- The most commonly used seeds are sunflower, poppy, sesame and pumpkin seeds.

Nutritional value of seeds

Seeds contain LBV protein, essential fatty acids, vitamins from the B group, Vitamin E, zinc and iron.

Storage of seeds

- Store seeds in airtight containers.
- Use seeds by their best-before date.

Exam tip

When an exam question asks for suggestions on how to make a recipe suitable for a vegan, you can suggest the use of a soya alternative. Remember to include details about how soya is a source of HBV protein so will directly replace an animal protein source and provide the same essential amino acids. Soya products tend to be bland, so include suggestions on how to marinade, or add extra flavour to the soya product of your choice.

Now test yourself

TESTED

1 Explain why soya milk is an acceptable alternative for someone who is lactose intolerant. Suggest a recipe where this could be substituted for cow's milk. [4 marks]
2 List four nutrients found in beans, and explain why they are important in our diet. [4 marks]
3 Name three different types of nuts that are used in cooking and provide a suggestion of where they can be used. [3 marks]

Fats and oils

Butter

- Butter is made by churning milk into a solid fat.
- It will be soft at room temperature.
- It is available in salted and unsalted varieties.
- Butter is used for creaming, when making cakes, spreading on bread, melting over vegetables, frying and shortening when making pastry.

Nutritional value of butter

Butter contains saturated fat, vitamins A and D and sodium (salt) in salted varieties.

Figure 1.13 Butter

Storage of butter

- Butter should be stored in the fridge, used by its use-by date and kept away from strong smelling foods.
- It can be frozen.
- If kept at room temperature, it should be in a container with a lid.
- It can go rancid if left for too long.

Oils

- Oils are liquid at room temperature.
- They are made from plant or seed sources.
- Examples include sunflower, rapeseed, olive and corn oil.
- They are used for frying, salad dressing, greasing or marinating.
- They can be flavoured with herbs, chilli or other flavours.

Nutritional value of oils

Oils contain unsaturated fats.

Storage of oils

Store away from direct sunlight in a cool place.

Margarine

- Margarine is a fat made from vegetable oils as an alternative for butter.
- Margarine has vitamins A and D added to it by law.
- It is available as a hard block or softened in a tub.
- It is used for baking or as a spread.
- Some spreads have a high water content and are not suitable for baking.

> **Exam tip**
>
> Make sure you know the main differences between fats and oils and are able to explain uses for them. If asked to give an example of an oil, give a named oil, such as olive or sunflower.

Nutritional value of margarine

Margarine contains a high amount of fat, vitamins A and D and sodium.

Storage of margarine

Store in the fridge and used by its sell-by date.

> **Now test yourself**
>
> TESTED
>
> 1 Give two uses for butter when cooking. [2 marks]
> 2 List two differences between fats and oils. [2 marks]
> 3 Identify one fat and one oil and give a use for each one in cooking. [4 marks]

Sugar and syrup

Sugar

- Sugar is made from sugar cane or sugar beet.
- It is used for sweetening drinks, sprinkling on cereal and making sweet baked foods.
- Sugar can add flavour and texture.
- It will aerate a product if creamed with a fat.
- It adds colour, as it caramelises when heated.
- Sugar also preserves a product, such as jam or jelly.
- It comes in many different forms:
  - White sugar – caster, granulated and icing sugars.
  - Brown sugar – Demerara, soft brown or dark brown sugar.

Nutritional value of sugar

Sugar provides 'empty calories', as it will be used as an energy source, but it provides no other nutrients.

Storage of sugar

- Sugar has a long shelf life if correctly stored.
- Store sugar in a cool, dry place and keep away from strong flavours and odours.

Figure 1.14 Different types of sugar

Syrup

- Golden syrup is used for sweet baked recipes and as a marinade.
- Black treacle is used in gingerbreads and dark fruitcakes.

Nutritional value of syrup

Syrup contains no nutrients but is an energy source.

Storage of syrup

- Store syrup in a cool, dry cupboard and use within three months of opening.
- Syrup can crystallise if left unused for a length of time.

Now test yourself

TESTED

1 Identify three types of sugar and state a recipe or dish where each type is used. [3 marks]
2 Explain how sugar acts as an aerator when making a creamed cake mixture. [4 marks]

> **Exam tip**
>
> A question on sugars is likely to ask about the function of sugar in a recipe. Check whether it is being used simply as a sweetener, or if it has other functions such as aeration or browning. Look at the number of marks available for the question to see whether a simple list will gain maximum marks, or if you need to provide a more detailed explanation of the function.

2 Macronutrients and micronutrients

The fuel and chemicals that we need for our bodies are called **nutrients**. Nutrients are divided into two main groups:

- **Macronutrients** (macro means large): needed these in large amounts. These nutrients are proteins, fats and carbohydrates.
- **Micronutrients** (micro means small): needed in small amounts. These nutrients are vitamins, minerals and trace elements.

The body also needs other substances in food to work properly, including water and fibre.

Macronutrients

Proteins

REVISED

- Proteins are very large molecules and are made of small units called **amino acids**.
- There are nine **essential amino acids**. These are the amino acids that cannot be made by our bodies, so we must eat the proteins that contain them.
- Foods that contain all of the essential amino acids are called **High Biological Value (HBV)** proteins.
- Foods that contain only some of the essential amino acids are called **Low Biological Value (LBV)** proteins.
- If we combine LBV proteins in a meal, we can provide all the essential amino acids for our bodies. This is called food combining or using **complementary proteins**. An example of a meal using complementary proteins is beans on toast or rice and dhal.

Figure 2.1 HBV sources of protein

Figure 2.2 LBV sources of protein

Table 2.1 Functions, sources and consequences of too little or too much protein

Functions of protein in the body	• Growth • Repair • Maintains the body • Produces enzymes for digestion, muscle activity and nerve function • Produces hormones to regulate body functions • A second source of energy	
Sources of protein	**HBV sources:** • Meat • Fish • Eggs • Milk • Cheese • Soya beans	**LBV sources:** • Cereals (rice, oats, quinoa, wheat, millet) • Peas, beans (except soya beans) and lentils • Nuts and seeds
What happens if we don't get enough protein?	Children: • Poor growth • Thinning hair, or hair loss • Cannot digest food properly so get diarrhoea • Catch infections easily • Low energy levels, lose weight and are thin and weak • Fluid under their skin (oedema)	Adults: • Lose fat, muscle and weight • Fluid under the skin (oedema) • Cuts and bruises may be slow to heal • Lack energy • Dry hair and skin • Catch infections more easily
What happens if we get too much protein?	• Puts strain on kidneys and liver • Increase in weight, as extra protein is converted into fat which is then stored in the body	

How much protein should we eat?

This depends on our age, our lifestyle and our activities.
- Babies, children and teenagers are still growing and therefore need more protein. Adults still need protein to help their hair and fingernails grow and for the body to repair.
- Pregnant women need protein to allow their baby to develop, and women who are breastfeeding (lactating) need protein to make their milk.

Nutritionists and scientists have worked out how much protein is needed by individuals. These are called the **Dietary Reference Values (DRVs)**.
- Babies and children up to the age of six years old: 12 to 20 grams of protein per day.
- Children seven years to 15 years: 28 to 42 grams of protein per day.
- Teenagers and adults over the age of 15: 45 to 55 grams of protein per day.

Exam tip

You may be asked to state or describe ways someone could ensure they are getting enough protein when they are a vegetarian and do not eat animal products. You need to list foods that contain LBV protein, and talk about combining foods to ensure all the essential amino acids are eaten, talking about complementary proteins and also giving some examples.

Now test yourself

TESTED

1 List three functions of protein in the diet. [3 marks]
2 Explain what an HBV protein is, giving two examples of where you find HBV proteins. [4 marks]
3 Why are cereals, nuts and seeds classed as LBV proteins? [2 marks]
4 List three symptoms you might see in an adult who is not eating enough protein. [3 marks]
5 Explain why someone may put on weight if they are eating too much protein. [3 marks]

Fats

- The general term for fats and oils is **lipids**.
- Fats are usually solid at room temperature. Oils are liquid at room temperature.
- Some fats are visible, such as the fat on meat, and butter or oils that we use for frying or salad dressing.
- Other fats are invisible and form part of a product that we eat, such as biscuits, ice cream or ready meals.

Figure 2.3 Visible fats

Figure 2.4 Invisible fats

Table 2.2 Functions, sources and consequences of too little or too much fat

Function of fats in the body	• Good energy source • Forms the structure of some cells • Insulates the body against the cold • Protects our vital organs, such as the liver and kidneys • A good source of vitamins D, E and K • Gives food texture and flavour • Helps fill us up	
Types of fats and their sources	**Saturated fats** Butter, lard, suet and animal fat found on meat **Unsaturated fats** There are two types: • **monounsaturated fat** sources: olive oil, avocados, nuts and oily fish like mackerel • **polyunsaturated** fat sources: seeds, oils, walnuts, salmon and sardines	**Essential fatty acids** • Omega 3: oily fish, seeds, walnut oil and green leafy vegetables • Omega 6: vegetables, fruits, grains, chicken and seeds
What happens if we don't get enough fat?	Babies and children: normal growth will be affected Adults: • Become thinner • Feel colder • Lack the essential fatty acids Omega 3 and Omega 6, which can put heart function and cholesterol levels at risk	
What happens if we eat too much fat?	• Weight gain • Extra fat may be stored in the liver and cause health problems • Saturated fat can also increase the risk of stroke and heart disease • Hydrogenated fats can increase the risk of cancer, diabetes, obesity and bone problems	

How much fat should we eat?

- Average man: no more than 95 g of fat per day, of which not more than 30 g should be saturated fat.
- Average woman: no more than 70 g of fat per day, of which not more than 20 g should be saturated.
- Children: about 35% total intake of food as fat. This ensures the child is getting sufficient energy sources and vitamins D, E and K.

> **Typical mistake**
>
> You may be asked to state or describe ways someone can reduce their fat intake. A common exam mistake is to write 'eat less' or 'add less when cooking' – you will gain limited marks. Instead, think about:
> - different cooking methods (for example, grilling rather than frying)
> - cutting off visible fat
> - choosing foods that are lower fat options (for example, leaner cuts of meat).

Now test yourself

TESTED

1 State three functions of fats in the body. [3 marks]
2 (a) Name three foods to avoid to reduce your intake of saturated fats. [3 marks]
  (b) Give two foods that you could eat instead to provide healthier sources of fat. [2 marks]
3 List four problems that may occur in the body if we eat too much fat. [4 marks]
4 Name two foods that will provide our bodies with Omega 3 fatty acids. [2 marks]
5 Explain the term 'invisible fat' and give an example of where this may be found. [3 marks]

Carbohydrates

REVISED

Carbohydrates are divided into simple carbohydrates and complex carbohydrates.

- Simple carbohydrates are **monosaccharides** or **disaccharides**. They are sugars, and are easily broken down by the body for energy.
- Complex carbohydrates are **polysaccharides**, which are large molecules that take a long time to digest.

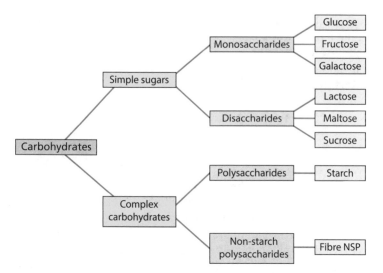

Figure 2.5 Simple and complex carbohydrates

Table 2.3 Functions, sources and consequences of too few or too many carbohydrates

Functions of carbohydrates in the body	Provide energy for physical activity and maintaining body functions • Sweeten and flavour foods • Provide fibre (NSP) to help digestion	
Sources of carbohydrates	**Monosaccharides** • Glucose: fruits and vegetables • Fructose: honey and some fruit and vegetable juices • Galactose: milk	**Disaccharides** • Sucrose: sugar cane and sugar beet – all types of sugar and also golden syrup and black treacle • Lactose: milk • Maltose: grains
	Polysaccharides • Starch: in grains, cereals, pasta and some fruit and vegetables • Soluble fibre: oats, beans, peas, lentils and most fruit and vegetables • Insoluble fibre: wholemeal bread and pasta, wholegrain cereals, brown rice and some fruit and vegetables	
What happens if we don't get enough carbohydrates?	• Weight loss • Body uses protein as a secondary energy source • Constipation • Increased risk of developing bowel cancer or diverticulitis	
What happens if we eat too many carbohydrates?	• Weight gain (excess carbohydrates are converted to fat and stored in your body) • Too much sugar can cause tooth decay	

How many carbohydrates should we eat?

- The Eatwell Guide advises that one third of your daily food should be starchy carbohydrates like bread, pasta, rice and potatoes.
- Starchy carbohydrates can also provide you with fibre if wholemeal and wholegrain products are consumed.
- An average (2,000 calories a day) diet should contain about 250 g of carbohydrates. (This would be, for example, a total of a bowl of wholegrain cereal, one portion of pasta and three slices of bread).

Exam tip

Some questions may ask about increasing the amount of carbohydrates in the diet to help with weight loss. In this case you would concentrate on polysaccharides, which take a long time to digest, and therefore fill you up for longer. A good example of this would be to encourage people to eat a starchy breakfast, such as porridge, wholegrain cereal or wholegrain muffins, flapjacks or wholemeal toast.

Now test yourself

TESTED

1 Name two monosaccharides and state where they are found in the food you eat. [4 marks]
2 Explain why the body takes longer to digest polysaccharides than disaccharides. [3 marks]
3 Explain how NSP helps the digestive system to work more efficiently and prevents constipation. [4 marks]
4 Discuss why eating a wholegrain breakfast cereal with milk can help someone who is on a weight-loss diet. [4 marks]
5 Explain why a diet that is high in sugary food is bad for young children. [4 marks]

Micronutrients

Vitamins

Vitamins are divided into two main groups: **fat-soluble vitamins** and **water-soluble vitamins**.

- Fat-soluble vitamins: when you cook foods containing these vitamins in fat, the vitamins will leach out of the foods into the fat.
- Water-soluble vitamins: when you cook foods containing these vitamins in water, the vitamins will leach out of the foods into the water.

Table 2.4 Fat-soluble vitamins

	Function in the body	Sources	What happens if we don't get enough?	What happens if we eat too much?
Vitamin A	<ul><li>Helps eyesight</li><li>Keeps the throat, digestive system and lungs moist</li><li>It is an antioxidant: helps stop damaging substances from the air and water entering our bodies</li></ul>	<ul><li>Found as retinol in animal foods such as milk, cheese, butter and oily fish, liver and liver products.</li><li>Found as beta-carotene in plant foods (for example, red and orange fruit and vegetables and darkgreen leafy vegetables)</li></ul>	<ul><li>Unable to see in dim light (called night blindness)</li><li>Difficult for the body to fight disease and infections</li><li>Children will have poor growth</li></ul>	<ul><li>Too much Vitamin A can be poisonous – it will build up in the liver</li><li>Pregnant women should avoid eating too much of the retinol as it can cause birth defects in the developing baby</li></ul>
Vitamin D	<ul><li>Controls the amount of calcium that is taken from food</li><li>Helps to develop strong bones and teeth</li><li>Has also recently been found to be important in brain function</li></ul>	<ul><li>Dairy products: butter, cheese, milk and eggs</li><li>Liver</li><li>Oily fish</li><li>Added by law to margarine</li><li>Most comes from exposure to sunlight</li></ul>	<ul><li>Weak bones and teeth</li><li>Young children will develop rickets</li><li>Older people develop osteomalacia</li><li>Elderly people develop osteoporosis</li></ul>	<ul><li>It is unlikely you will eat too much Vitamin D in your diet</li><li>Taking Vitamin D supplements over a long period of time can cause more calcium to be absorbed in your body, which may damage the kidneys</li><li>Excessive intake can also weaken bones</li></ul>
Vitamin E	<ul><li>It is an antioxidant</li><li>Helps keep the cell walls in the body healthy</li><li>Helps prevent the development of some cancers and heart disease</li></ul>	<ul><li>Vegetable oils</li><li>Lettuce</li><li>Peanuts</li><li>Seeds</li><li>Wheat germ oil</li></ul>	<ul><li>Deficiency is very rare; occasionally it can arise if a person has a problem absorbing fat</li></ul>	<ul><li>It is unlikely you will eat too much Vitamin E from a normal diet</li></ul>
Vitamin K	<ul><li>Helps the blood to clot</li></ul>	<ul><li>Green leafy vegetables</li><li>Cheese</li><li>Liver</li><li>Asparagus</li><li>Coffee</li><li>Bacon</li><li>Green tea</li></ul>	<ul><li>It is very rare to have a deficiency in adults</li></ul>	<ul><li>It is very unlikely that you will eat too much and there is no evidence that too much will harm you</li></ul>

Table 2.5 Water-soluble vitamins

	Function in the body	Sources	What happens if we don't get enough?	What happens if we eat too much?
Vitamin B$_1$ (thiamin)	<ul><li>Helps release energy from carbohydrates</li><li>Helps your nerves to work properly</li><li>Helps with growth in the body</li></ul>	<ul><li>Wheat and rice</li><li>Cereal products and wheat germ</li><li>Yeast and marmite</li><li>All types of meat</li><li>Eggs and fish eggs (roe)</li><li>Milk and dairy food</li><li>Seeds, nuts and beans</li></ul>	<ul><li>Slow growth and development</li><li>Severe deficiency causes beri-beri, which causes muscle wastage</li></ul>	<ul><li>There is currently no evidence to suggest that eating too much will cause any harm</li></ul>
Vitamin B$_2$ (riboflavin)	<ul><li>Helps release energy from carbohydrates</li><li>Helps your growth and keeps your skin healthy</li></ul>	<ul><li>Liver and kidneys</li><li>Meats</li><li>Eggs and milk</li><li>Green vegetables</li></ul>	<ul><li>Dryness of the skin around the mouth</li><li>Poor growth</li></ul>	<ul><li>There is currently no evidence to suggest that eating too much will cause any harm</li></ul>
Vitamin B$_3$ (niacin)	<ul><li>Helps release energy from carbohydrates</li><li>Essential for healthy skin and nerves</li><li>Can help lower levels of fat in the blood</li></ul>	<ul><li>Meat and poultry</li><li>Cereals and grains</li><li>Dairy products</li><li>Pulse vegetables (for example, lentils)</li></ul>	<ul><li>You can develop a disease known as pellagra, which can cause diarrhoea, dermatitis and dementia</li></ul>	<ul><li>It is unlikely you will eat too much Vitamin B$_3$ in a normal diet</li></ul>
Vitamin B$_6$ (pyridoxine)	<ul><li>Helps release energy from carbohydrates</li></ul>	<ul><li>Found in a wide range of foods</li></ul>	<ul><li>Headaches</li><li>General aching and weakness</li><li>Anaemia</li><li>Skin problems</li></ul>	<ul><li>Taking large amounts of Vitamin B$_6$ in supplements can lead to a loss of feeling in the arms and legs known as peripheral neuropathy</li></ul>
Vitamin B$_9$ (folate or folic acid)	<ul><li>Helps the body to use proteins</li><li>Essential for the formation of DNA in the body cells, especially the cells that make red blood cells</li></ul>	<ul><li>Liver and kidneys</li><li>Wholegrain cereals</li><li>Pulses</li><li>Darkgreen vegetables</li></ul>	<ul><li>Tiredness</li><li>Anaemia</li><li>Lack of folate in pregnancy can cause the foetus to have spinal malformations and spina bifida</li></ul>	<ul><li>Taking large doses of folic acid can disguise a Vitamin B$_{12}$ deficiency, which can be a problem with older people</li></ul>
Vitamin B$_{12}$ (cobalamin)	<ul><li>Needed to form a protective coating around nerve cells to make them work properly</li><li>Important for the production of new cells</li></ul>	<ul><li>Meat, fish and eggs</li><li>Dairy products</li><li>Vegans need to take a supplement as it is only found in animal foods</li></ul>	<ul><li>Tiredness and anaemia</li><li>Muscle weakness, 'pins and needles'</li><li>Depression</li><li>Memory problems</li></ul>	<ul><li>There is currently no evidence to suggest that eating too much will cause any harm</li></ul>

	Function in the body	Sources	What happens if we don't get enough?	What happens if we eat too much?
Vitamin C (ascorbic acid)	• Helps with absorption of iron from other foods • Produces collagen, which makes connective tissues to bind cells together in the body • It is an antioxidant, which means it helps protect the body against polluting chemicals that harm us	• Fruits, especially citrus fruits (for example, oranges and lemons), blackcurrants and kiwi fruits • Tomatoes • Green leafy vegetables (not lettuce) • Peas • New potatoes • Broccoli	• Rare, but if you do not eat enough fruit and vegetables it can happen • Slight deficiency can cause anaemia • Severe deficiency leads to scurvy, which means you will have tiredness, bleeding gums and anaemia	• Excess Vitamin C eaten in a normal food source is excreted by the body • Excess Vitamin C taken as a supplement can cause nausea and diarrhoea

Figure 2.6 Oranges are a good source of Vitamin C

> **Exam tip**
>
> Questions on vitamins often ask how you can prevent loss of Vitamin C during cooking. You need to remember that Vitamin C is water-soluble and suggest ways to cook that use a minimum of water, for example steaming, microwaving and stir-frying.

Now test yourself

TESTED

1. Identify two problems caused by a lack of Vitamin A and state what will happen to the body in each case. [4 marks]
2. Name two sources of foods that contain folic acid and explain why it is important that pregnant women eat enough folic acid. [4 marks]
3. Explain why a vegan would need to take supplements containing Vitamin B$_{12}$ and state what will happen if they do not take these supplements. [3 marks]
4. Explain why it is important to eat foods containing Vitamin C with iron-rich foods, particularly if you only eat foods containing non-haem iron. [4 marks]

Minerals and trace elements

REVISED

● **Minerals** help to make strong bones and teeth, make sure we have sufficient red blood cells to transport oxygen around the body, control the amount of water in our body and make the nerves and muscles work correctly. They are needed in quantities of between 1 mg and 100 mg per day.

● **Trace elements** are responsible for strengthening the tooth enamel, making hormones and controlling bodily functions, and act in other muscle and nerve functions. They are needed in minute quantities, less than 1 mg per day.

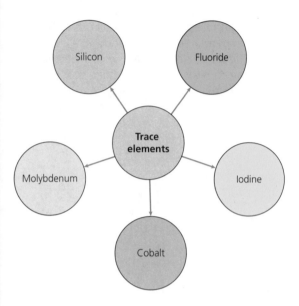

Figure 2.7 Trace elements

Table 2.6 Functions, sources and consequences of too little or too much of each mineral

	Function in the body	Sources	What happens if we don't get enough?	What happens if we eat too much?
Calcium	● Works with phosphorus and Vitamin D to make strong, healthy bones and teeth ● Helps with blood clotting ● Involved with nerve and muscle function	● Dairy products ● Added to white bread by law ● Oily fish ● Green vegetables ● Nuts/seeds ● Citrus fruits ● Soya milk ● Fruit juices and yoghurts may have calcium added	● Poor bone structure ● Peak bone mass will not be reached, causing weak bones or osteoporosis later in life ● Babies' bones will not form correctly ● Blood will not clot properly after an injury	● It is unlikely you will eat too much calcium ● Taking supplements of over 1,500 mg per day can cause stomach pain, diarrhoea and constipation
Iron	● Needed to make haemoglobin, which is the red-coloured protein in red blood cells that transports oxygen around the body	● Red meat, kidneys and liver ● Egg yolk ● Green leafy vegetables ● Dried apricots ● Lentils ● Cocoa and plain chocolate ● Some bread and cereals are fortified with iron	● Anaemia ● A pale complexion ● Pale inner eyelids ● Weak and split fingernails	● The side effects of eating more than 20 mg of iron per day are stomach pains, nausea, vomiting and constipation

	Function in the body	Sources	What happens if we don't get enough?	What happens if we eat too much?
Magnesium	• Supports a healthy immune system • Prevents inflammation • Involved in the digestive process	• Green leafy vegetables • Nuts • Brown rice • Wholegrain bread • Fish, meat and dairy foods	• Weak muscles • Abnormal heart rhythm • A slight rise in blood pressure	• Too much naturally occurring magnesium has not been shown to cause any side effects
Potassium	• Helps build proteins • Breaks down carbohydrates • Builds muscle • Controls the electrical activity of the heart • Maintains normal body growth	• All red meats • Fish (for example, salmon and sardines) • Broccoli, peas, sweet potatoes and tomatoes • Bananas, kiwi fruit, dried apricots • Milk and yoghurt • Nuts	• Weak muscles • Abnormal heart rhythm • A slight rise in blood pressure	• It is extremely unlikely you will eat too much potassium in a normal diet
Sodium	• Maintains water balance in the body • Involved in nerve transmissions	• Cheese • Bacon • Smoked meats • Processed foods • Table salt (sodium chloride) • Monosodium glutamate and sodium bicarbonate are both additives and contain sodium	• Deficiency is unlikely as it is in so many foods • If you are exercising in hot conditions you may get muscle cramps • You can lose sodium if you have sickness and diarrhoea	• High blood pressure • Fluid retention • Risk of heart failure and stroke • Damages kidneys in young babies and children

Exam tip

Exam questions on minerals usually focus on calcium, iron and sodium.
- Calcium questions can be about the effects of not eating enough calcium, particularly if you are lactose intolerant. You may have to suggest different ways of someone getting calcium in their diet other than eating dairy foods.
- Questions on iron could concentrate on how a vegetarian or vegan can ensure they are getting enough iron. You could discuss haem and non-haem iron, giving examples of foods containing both sorts. You could then emphasise the importance of eating foods rich in Vitamin C, which helps with the absorption of non-haem iron.
- Questions on sodium could ask about reducing sodium or salt levels in the diet. You would need to discuss what a diet high in sodium can do to the body, where sodium is found and suggest ways to reduce this amount in the diet.

Now test yourself

TESTED

1 Suggest three ways a vegan could increase the amount of iron in their diet. [3 marks]
2 List three alternative flavourings someone could use instead of salt if they are trying to reduce the amount of salt in their diet. [3 marks]
3 (a) Explain the symptoms of anaemia. [2 marks]
  (b) Give one reason that young women are more likely to suffer with lack of iron than young men. [1 mark]
4 What is osteoporosis? List ways that someone can help to reduce the risk of developing this condition. [4 marks]

Water

- is in all body cells and is involved in chemical reactions in the body
- regulates the body's temperature, keeping it around 37 °C
- is found in all body fluids, such as blood, urine, saliva, digestive juices and sweat
- helps get rid of waste products in faeces and urine
- keeps the linings of our digestive system, lungs, nose and throat moist
- helps us absorb nutrients
- transports nutrients, oxygen and carbon dioxide around the body in the blood.

Figure 2.8 Water is a vital part of our diet

We obtain plenty of water in the foods that we consume, particularly fruit and vegetables.
- We should be taking in 1.75–2 litres of water a day in drinks and food. All fluids that contain water count, including tea and coffee.
- Sweet and sugary drinks are likely to make you more thirsty, as the sugar will quickly enter the bloodstream and make the body need water to dilute this down.

If we do not drink sufficient water we can become **dehydrated**. The symptoms of dehydration are:
- headache
- dark-coloured urine
- weakness and nausea
- overheating of the body
- confusion
- changes in blood pressure.

If you drink too much water, your kidneys will not be able to cope and your blood will become diluted. Your brain will swell and this can cause nausea, convulsions and possibly death.

Exam tip

Questions on water intake are likely to focus on what may happen if we do not drink sufficient water. Make sure you include reference to fresh fruit and vegetables containing large amounts of water.

Now test yourself

TESTED

1 State three functions of water. [3 marks]
2 List four symptoms that may be apparent from not drinking enough water. [4 marks]

Dietary fibre (Non-starch polysaccharide – NSP)

Fibre is a **polysaccharide**. This means it is a very complex carbohydrate. Our bodies cannot digest it, so it provides bulk in the diet and helps to move the waste food through the system, preventing constipation and cleaning the walls of the digestive system to remove bacteria.

Functions of NSP

REVISED

- Helps prevent bowel disorders including bowel cancer, diverticular disease and haemorrhoids (piles).
- Helps with weight control as high fibre foods are filling, but the fibre is not digested.
- High fibre diets have been shown to help lower blood cholesterol.

Soluble fibre:
- slows down the digestive process and the absorption of carbohydrates, so makes us feel full for longer
- helps control blood sugar levels
- helps to lower blood cholesterol levels.

Good sources of soluble fibre are oats, beans, peas, lentils and most fruit and vegetables, particularly if you eat the skin.

Insoluble fibre absorbs water and increases bulk, so keeps faeces soft, making them pass through the digestive system easily. This prevents constipation.

Good sources of insoluble fibre are wholemeal bread and pasta, wholegrain cereals, brown rice and some fruit and vegetables.
- We need a minimum of 18 g of fibre per day, but the ideal amount is 30 g per day. Most people in the UK do not eat enough fibre.
- Children need less fibre, as it will fill them up too quickly and may mean they are not getting enough of the other nutrients needed for their healthy growth.

> **Exam tip**
>
> Questions on fibre are likely to ask ways that you can increase the fibre intake. Remember to list them, explaining how this can increase the intake.
>
> An example could be: change white bread to wholemeal bread, as the wholemeal flour used to make the bread contains the outer husk of the wheat grain, which adds fibre to the product.

Now test yourself

TESTED

1 List three health problems that can result from a diet low in fibre. [3 marks]
2 You are making a steak and kidney pie for someone who is lacking fibre but does not like vegetables. How can you increase the fibre content of the dish? [3 marks]
3 Discuss ways you can encourage a young person to increase the amount of fibre in their diet. [4 marks]

How much energy does each of the macronutrients provide?

- 1 gram of carbohydrates provide 3.75 kilocalories.
- 1 gram of pure fat provides 9 kilocalories.
- 1 gram of pure protein provides 4 kilocalories.

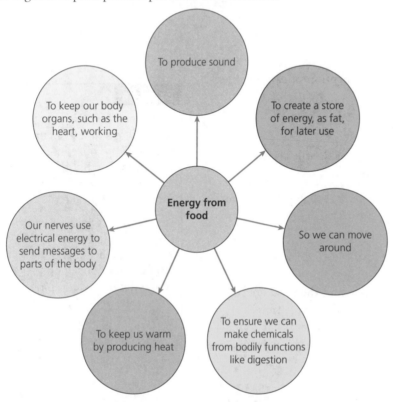

Figure 3.1 Uses of energy

Vitamins and minerals are not broken down by the body so do not produce energy.

Recommended Daily Intake (RDI) and percentage energy values of nutrients

REVISED

Table 3.1 Recommended Daily Intake (RDI) and percentage energy values of nutrients

Nutrient	Total amount in a 2,000 kilocalorie per day diet for an adult	Percentage of energy from this nutrient
Total fat	70 grams	35%
Of which saturated fat:	20 grams	11%
Total carbohydrates	260 grams	50%
Of which sugars:	50 grams	5% from extrinsic sugars
		45% from intrinsic sugars and starches
Protein	50 grams	15%

- **Extrinsic sugars** are those you can see (such as the sugar you put into cake and biscuit recipes, or those added to sugary, fizzy drinks).
- **Intrinsic sugars** are those you cannot see (such as the sugar in fruit)

The total amount of extrinsic sugar consumed per day should be around 7 teaspoons. A regular 330 ml can of fizzy drink such as Coca Cola has about 6 teaspoons of sugar in it.

> **Exam tip**
>
> You may be asked how people can cut down on their calorie intake. You need to include information about how many kilocalories are provided by each of the three macronutrients, and show that by cutting down on fats, a person can reduce their overall calorie intake. Include information about the guidelines – the RDIs recommended by the government for maximum intake in a 2,000 calorie per day diet.

Now test yourself

TESTED

1 There are concerns about the number of children who have significant problems with dental caries. Suggest three reasons why this is happening, making reference to the amount of extrinsic sugars being consumed. [4 marks]
2 Give two reasons why the government recommend that only 20 g of saturated fat should be consumed daily in a 2,000 kilogram per day diet for an adult. [2 marks]
3 Discuss why it is better for someone trying to lose weight to eat more carbohydrates than fatty foods, and suggest two dishes that they could eat for breakfast to help with weight loss. [6 marks]

How do nutrients work together in the body?

Complementary actions of nutrients

REVISED

The way nutrients work together is called **complementary action**.

Table 3.2 Complementary nutrients

Which nutrients work together?	What do they do together?
Vitamin D and calcium	Vitamin D helps the body absorb calcium.
Vitamin C and iron	Vitamin C helps the body to take up the plant-based non-haem iron that is found in vegetables and is more difficult for the body to absorb.
Sodium and potassium	Eating foods rich in potassium helps encourage the kidneys to get rid of excess sodium. Too much sodium can contribute to high blood pressure, stroke and heart attacks.
Niacin and tryptophan	Niacin (Vitamin B_3) and tryptophan work together to build new protein in the body.
Vitamin B_{12} and folate/folic acid (vitamin B_9)	These work together to help with cell division and replication, during foetal development and healing processes.
Zinc and copper	These compete with each other to be absorbed in the intestine. Try to avoid eating foods that contain these trace elements at the same time.
Dark green leafy vegetables and fats	Most of the minerals in these vegetables are fat-soluble, so eating a healthy source of fat with greens will increase the absorption of these minerals in the body.

Figure 3.2 Cereal fortified with Vitamin D with milk

Now test yourself

TESTED

1 Suggest two meals that a vegetarian could eat that would provide both Vitamin C and iron and explain why it is important that these foods are eaten together. [4 marks]
2 Explain why it is important that foods containing Vitamin D and calcium are eaten together. [3 marks]
3 Suggest two ways in which dark green leafy vegetables can be cooked to help with the absorption of the minerals in these. [2 marks]

How much energy do we need?

Basal metabolic rate (BMR)

REVISED

The amount of energy needed just to keep everything working to stay alive is called the **Basal Metabolic Rate (BMR)**. This is the energy required to:

● keep breathing
● make chemicals in the body
● keep your heart beating
● keep other body organs working
● keep your blood pumping and your nerves working.

Table 3.3 Basal Metabolic Rates (kilocalories per day)

Boys (1–17 years old)	Girls (1–17 years old)	Men (18–75 years old)	Women (18–75 years old)
550–1,500	500–1,350	1,695–1,350	1,350–1,090
kilocalories	kilocalories	kilocalories	kilocalories

● The BMR of a child will increase as they grow.
● The BMR of adults reduces slightly as they increase in age.

Physical activity level (PAL)

REVISED

Your **Physical Activity Level (PAL)** is the amount of physical activities you do, like sitting, standing, walking, running or planned and structured exercise. This will change depending on how active you are.

Table 3.4 Physical Activity Levels

PAL	Daily activities	Lifestyle
Less than 1.4	Hospital patient, in bed	Inactive
1.4 to 1.55	Little physical activity at work or in leisure time	Sedentary
1.6	Moderate physical exercise, female	Moderately active
1.7	Moderate physical exercise, male	Moderately active
1.7 to 2.0	Construction worker; someone who works out at the gym for an hour a day	Moderately active
2.0 to 2.4	Physically active at work, e.g. fitness trainer	Very active
2.4+	Professional athlete, footballer, etc.	Extremely active

Figure 3.3 You can increase your PAL by being more active

Estimated Average Requirement (EAR)

REVISED

To calculate how much energy you need to consume to maintain your weight, a calculation is made to find your **Estimated Average Requirement (EAR)**, and to calculate the number of calories you need per day.

**Basic Metabolic Rate × Physical Activity Level
= Estimated Average Requirement**

BMR x PAL = EAR

- The EAR of children will increase between birth and the age of 18 years.
- The EAR for girls is less than for boys.
- The EAR for adults will increase as the Physical Activity Level (PAL) increases.
- The EAR for adults decreases with age as people become less active.

> **Exam tip**
>
> Questions on BMR and EAR will sometimes give a described situation for a given individual and ask questions about how they can increase their PAL to help lose weight. For example, an office worker may not do any exercise during the week, and as they sit down for most of the day they need to increase their PAL. You would need to make suggestions as to how they could do this.

Now test yourself

TESTED

1 Why does the Basic Metabolic Rate (BMR) of a child increase as they grow? [2 marks]
2 Explain why a professional athlete has a much higher Estimated Average Requirement (EAR) than a bus driver. [4 marks]
3 Give two reasons why the EAR for girls is less than for boys. [2 marks]
4 Suggest three ways that a teenage girl could increase their Physical Activity Level (PAL). [3 marks]

4 Plan balanced diets

Recommended guidelines for a healthy diet

The Eatwell Guide

The Eatwell Guide shows the proportion of foods we should have in our diet to get the correct balance of healthy food. It shows how much of what we eat should come from each food group.

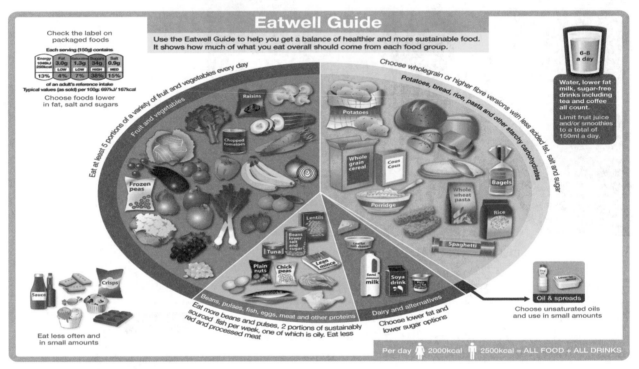

Figure 4.1 The Eatwell Guide

- Green segment: fruit and vegetables (a minimum of five portions per day)
- Yellow segment: starchy foods (meals should be based on these)
- Pink segment: protein foods (eat at least two portions of fish per week, one of which should be oily fish; eat more beans and pulses; reduce your intake of processed meats such as ham and bacon)
- Blue segment: dairy and alternatives
- Purple segment: unsaturated oils (such as vegetable/olive oil) and lower fat spreads (high in fat and calories, so only consume in small amounts)
- Foods high in fat and/or sugar: not included on the plate, as they are not part of a balanced diet
- Water (including tea, coffee and limited fruit juice): for hydration
- An example of the traffic light labelling system is included for information.

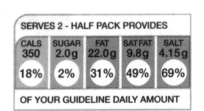

Figure 4.2 Traffic light labelling system

Recommended Daily Intake

The Government also recommends how much of each nutrient you should be eating. This is called the **Recommended Daily Intake (RDI)**.

Table 4.1 RDI for different age groups

Age	RDI (in grams)		
	Protein	Fat	Carbohydrates
1–3 years	15 g		
4–5 years	20 g		
6–10 years	28 g	70 g	220 g
11–14 years	42 g	70 g	220 g
15–18 years	55 g	70 g	230–300 g
Women	45 g	70 g	230 g
Men	55 g	95 g	300 g

There are no Recommended Daily Intake amounts of fat and carbohydrate for young children.

Eight tips for healthy eating

There are also eight healthy steps to eating a balanced diet:

1 **Base your meals on starchy foods:** These fill you up and provide slow-release energy.
2 **Eat a minimum of five portions of fruit and vegetables every day:** 80 g of fruit or vegetables, 3 tablespoons of vegetables, or a 150 ml glass of fruit juice all count as one portion. Fresh, frozen or canned fruit and vegetables all count. Each fruit or vegetable only counts once, no matter how many portions of that fruit or vegetable you eat per day.
3 **Eat at least two portions of fish a week, one of which should be oily fish:** white fish is low in fat; oily fish provides Omega 3 fatty acids for your heart health.
4 **Cut down on saturated fat and sugar:** This will help prevent heart disease, obesity and dental caries. Look out for hidden fat in ready-made products.
5 **Eat less salt**, no more than six grams a day for adults: Eating too much salt can raise your blood pressure and cause heart disease. Some foods have hidden salt; look out for these.
6 **Drink plenty of water:** This will hydrate the body, help with digestion and help prevent constipation.
7 **Do not skip breakfast:** Research has shown that people who eat breakfast perform better at school or work, and are able to concentrate for longer.
8 **Get active and try to be a healthy weight:** People who exercise regularly are less likely to have health problems such as obesity, heart problems and loss of bone strength as they get older.

Figure 4.3 A portion of all of these counts towards your recommended five a day.

Food and Nutrition Strategy for Wales

The Food Standards Agency nutrition strategy **'Food and Well Being'** aims to improve the diet and health of the population of Wales, particularly young people, children and infants, low income and vulnerable groups

(including the elderly and ethnic minority groups), middle-aged men and women of childbearing age. The strategy recommends:
- increasing the uptake of a healthy, balanced diet
- increasing fruit and vegetable intake
- developing initiatives to prevent and manage obesity
- improving healthy eating using national schemes and policies
- providing information to key decision makers to tackle poor nutrition
- ensuring the public are well informed
- implementing local initiatives
- developing and promoting initiatives with the food industry.

Other initiatives include:
- **Food and Fitness – Promoting Healthy Eating and Physical Activity for Children and Young People in Wales** includes suggestions for how children and young people can increase physical activity levels. It also suggests improving provision of healthy food in schools, providing cookery and nutrition lessons and increasing the range of physical activities. It includes the development of the Welsh Network of Healthy School Schemes (WNHSS).
- **Change4Life/Newid am Oes:** Aims to help people achieve and maintain a healthy body weight through better nutrition and more exercise.
- **Healthy Working Wales:** A Welsh Government programme set up to support working age people in Wales to stay fit and healthy.
- **Health Challenge Wales:** Created by Public Health Wales.
- **Mind, Exercise, Nutrition, Do it!:** A referral plan for overweight and obese children.
- **Appetite for Life Action Plan:** Aims to improve the nutritional standards of food and drink offered in schools.
- **The Healthy Eating in Schools (Nutritional Standards & Requirements) (Wales) Regulations 2013:** These state the type of food that can and can't be provided by schools.

> **Typical mistake**
>
> Questions on a balanced diet may ask about the Eatwell Guide, asking you to describe why each section is a certain size. You need to explain which foods are in that particular section of the plate, and why the nutrients in that section are more important to the body than the smaller sections. Include information about the nutrients in the section and what their function is in the diet.

Now test yourself

TESTED

1 Explain why starchy foods is one of the largest sections on the Eatwell Guide. [4 marks]
2 List three of the eight recommended steps for a healthy diet as issued by the British Government and explain how following these three steps can improve your diet. [3 marks]
3 Why is it recommended that you eat two portions of fish per week, one of which should be oily fish? [2 marks]

How our nutritional needs change due to age

Babies and small children

REVISED

- From birth, babies are fed milk, either breast milk or bottle formulas.
- Once the baby is old enough, he or she is introduced gradually to solid food. This is called **weaning**.
- Foods that may cause allergies may be introduced gradually. If there is a known allergy to these in the immediate family, it is best to wait until the baby is over a year old.
- A good variety of foods should be introduced to ensure a balanced diet.
- Babies do not need sweet foods or added sugar, as this may create cravings later in life.

Children aged 1–4 years old

REVISED

- These children are usually very active and are growing quickly.
- They need small, regular meals and drinks to give them energy throughout the day.
- They also need a diet higher in fat to provide energy.
- Children this age need all the macro- and micronutrients as they are still growing, and their bodies are still forming.

Children aged 5–12 years old

REVISED

- This is an age where children should be very active and are growing rapidly.
- Exercise helps the bones to become stronger and take up the calcium they need. This is laying down bone density, creating peak bone mass, which will help prevent problems such as weaker bones in old age.
- Children at this age are advised to eat 28 grams of protein per day and around 1,900 kilocalories a day for boys and 1,700 kilocalories a day for girls.

Teenagers

REVISED

- This is an age where you are changing from a child into an adult.
- Girls will have growth spurts and will often start puberty earlier than boys. Teenagers can grow several centimetres in a few months.
- Boys will put on large amounts of muscle tissue and therefore sufficient protein is needed to support this growth.
- All the essential vitamins and minerals are needed to help the bones and internal organs form correctly. Peak bone mass is not reached until about 30 years of age.
- Teenage girls may be prone to being anaemic, as they will start their periods during this time. Iron-rich foods should be eaten, along with foods high in Vitamin C to help with iron absorption.
- Teenagers, particularly girls, are very conscious of body image.

Figure 4.4 Teenagers need regular, balanced meals

Adults and older people

REVISED

- Adults' bodies have stopped growing but they still need all the nutrients to keep the body working properly, and to repair and renew cells.
- Adults also need to prevent diseases and dietary-related conditions developing.

Early adulthood

- Calcium-rich foods should be consumed to continue laying down bone density to prevent problems with weaker bones in later life. (Peak bone mass is not reached until around the age of 30.)

Pregnancy

- Pregnant mothers need to eat a healthy, balanced diet, with extra care taken that they eat enough of the following nutrients:
  - Calcium is needed for the baby's skeleton. Most of the calcium is needed in the last three months of pregnancy. The mother has to ensure she has enough calcium for herself and the baby.
  - Vitamin D is needed to help the absorption of calcium in the body. If insufficient calcium is eaten, the baby will absorb calcium from the mother's bones, and these will become weaker.

> **Exam tip**
>
> Questions on this topic could ask you to design a meal for a person of a certain age group. Think about all the nutrients they require at that age, and make sure you include foods that will provide each nutrient. Explain why you are including a particular food, which nutrient it contains and why it is necessary for that particular age group.

○ Iron forms haemoglobin in the blood to carry oxygen during respiration. The baby has to build up a store of iron to last for the first three months of its life, as it will get no iron from the milk it is fed.

○ Vitamin C will help with iron absorption in the body.

○ Folic acid/folate will reduce the risk of the baby developing spinal defects, such as spina bifida.

○ Fibre is important because constipation may be a problem during pregnancy.

○ Avoid too many fatty and sugary foods and eat a balanced, healthy diet.

Middle adulthood

● Middle adulthood can be a time when people begin to be less active, so start to gain weight as their EAR (Estimated Average Requirement) reduces due to a decrease in PAL (Physical Activity Levels).

● It is important to make sure that you are eating the correct number of kilocalories to reduce the chances of developing diseases that are related to being overweight, such as diabetes.

● Middle-aged men are particularly vulnerable to coronary heart disease, and are a priority group for diet improvement and awareness.

Late adulthood

● It is important to keep the body weight within a healthy range. Older people have a lower BMR, and usually exercise less, so need fewer calories.

● Older people need to eat:

○ Vitamin D and calcium to help prevent osteoporosis

○ fibre and water to prevent constipation

○ Vitamin C and iron-rich foods to help with the absorption of iron to prevent anaemia

○ antioxidants to help prevent eye problems and cancers

○ foods containing Vitamin B_{12}.

● Older people need to avoid:

○ fatty and sugary foods, as they are less active and more likely to gain weight

○ sodium (salt), as this could lead to high blood pressure and heart disease.

● Other health problems could include difficulty chewing due to dentures or Parkinson's disease and difficulty cutting food up due to arthritis.

Now test yourself

TESTED

1 Explain why children aged between one and four years old should be given small, regular meals throughout the day. [2 marks]

2 Name two types of foods that older people should avoid, and explain why it is necessary for them to avoid these. [4 marks]

3 Explain the term of peak bone mass, and state why it is important that this is reached before the age of 30. [3 marks]

How our nutritional needs change due to lifestyle choices

Religious beliefs

REVISED

Hindus

● Many Hindus are vegetarian, but some eat fish.

● Even if a Hindu is not vegetarian, they do not eat beef or pork.

● Some do not eat eggs.

Muslims

- All food must be **Halal**, which means that meat is slaughtered in a special way.
- Unlawful or **Haram** foods include any pork or pork product, gelatine from a non-Halal-slaughtered animal, alcohol (or any foods containing alcohol), foods that contain emulsifiers made from animal fats, some types of margarine, drinks containing caffeine, and breads containing dried yeasts.
- Muslims have a special event called Ramadan. This is a fasting period in the ninth month of the Islamic calendar when no food is eaten during daylight.

Jews

- Jews must only eat **Kosher** food.
- Jews can only eat fish that have scales and fins.
- Pork and shellfish are forbidden.
- Dairy foods and meat must not be prepared, cooked or eaten together.

Exam tip

Questions on different religions often ask you to list rules about the diet of that religion, and then plan a meal for someone who follows that religion. Learn the basic rules for the three religions included here, and then think about dishes you have made that do not contain any banned foods for that religion.

Now test yourself

TESTED

1 Name two foods that Muslims are forbidden to eat. [2 marks]
2 Explain what Kosher food is. [3 marks]

Vegetarianism

REVISED

People become vegetarians for many reasons. For some people it is an ethical consideration, as they do not agree with killing animals for meat, or the way animals are treated during the rearing and killing processes. Other vegetarians do not like the taste of meat, or think a vegetarian diet is a healthier option. Some religions have a vegetarian diet.

Types of vegetarians

- **Lacto-ovo vegetarians:** do not eat meat or fish, but they will eat eggs, cheese and milk, and milk products such as yoghurts.
- **Lacto vegetarians:** eat the same as lacto-ovo vegetarians, but they do not eat eggs.
- **Vegans (sometimes called strict vegetarians):** do not eat meat or fish, or any animal products at all. This means no milk, cheese or eggs. They will eat only plant-based foods.

Figure 4.5 Vegans do not eat any animal products, including milk, eggs and cheese

What nutrients can they lack?

- **Iron:** Vegans only eat non-haem iron sources (plant sources), so need to eat foods containing Vitamin C with their meals.
- **Protein:** Vegans do not eat HBV protein, so complementary LBV protein sources should be eaten together.
- **Vitamin B$_{12}$:** This is found in animal foods only, so vegans should take supplements.

Exam tip

Questions on vegetarians can ask you to adapt a standard recipe to make it suitable for a vegetarian or vegan. Make sure you know which sort of vegetarian the question means. If it is a vegan, remember they do not eat any animal products, so you cannot use products that contain egg white.

Now test yourself

TESTED

1 List two nutrients that a vegan could be lacking in their diet. Explain what these nutrients do in the body and identify foods that the vegan could eat to obtain these nutrients. [6 marks]
2 Identify three reasons that someone might choose to become a vegetarian. [3 marks]
3 Explain the difference between a lacto-ovo vegetarian and a vegan. [4 marks]

Planning a balanced diet for people with specific dietary needs or nutritional deficiencies

Coeliac disease

- People with this condition have an intolerance of a protein called gluten. The lining of the small intestine is damaged by the gluten. This means that the absorption of nutrients becomes more difficult.
- **Coeliac disease** cannot be cured, but can be improved by eating a diet that is gluten free.
- Gluten free products are available in supermarkets.

Figure 4.6 Gluten-free products

Type 2 diabetes

- This is called **non-insulin dependent** diabetes and often develops later in life.
- Sugar-rich diets mean more young people, teenagers and children are being diagnosed with the condition.
- The blood sugar levels in the body are controlled by a hormone called insulin, which is released from the pancreas. Consumption of sugary foods causes the pancreas to stop releasing insulin.
- People who are overweight or obese are also at risk of developing this disease.
- If glucose stays in the blood it can eventually damage blood vessels in your eyes and cause blindness.
- It can restrict blood flow to your hands, feet and toes, resulting in infection and in the worst cases, amputation.
- It also can cause kidney damage.
- Type 2 diabetes can be controlled by eating a balanced, healthy diet based on complex, starchy carbohydrates.
- In 2015 there were 177,000 people in Wales with diabetes, and as obesity rates are higher in Wales than anywhere else in the UK, there is a much greater risk of developing Type 2 diabetes.

Cardiovascular disease

- **Cardiovascular disease** is a disease of the heart or blood vessels. Cardiovascular disease includes coronary heart disease and stroke.
- Blood flow to the heart can be restricted by a build up of fatty deposits on the walls of the arteries that supply the heart.
- This causes the arteries to harden and narrow.

Coronary heart disease (CHD)

- This is when the arteries supplying the heart become reduced in diameter, or blocked, and the blood cannot flow properly to the heart.
- This condition is caused by high levels of cholesterol in the blood.
- Other factors that contribute to CHD include high blood pressure and being overweight or obese, which can put a strain on the arteries and heart.
- You can lower the risk of developing CHD by:
  - eating more fruit, vegetables and fibre
  - selecting lower-fat foods

- ○ grilling or baking foods instead of frying
- ○ exercising, losing weight and stopping smoking
- ○ reducing salt intake.
- CHD is the main cause of death in Wales.
- Rates of heart disease are much higher in Wales than in England and many other European countries.

Stroke

- This happens when a blood vessel to the brain becomes blocked and part of the brain does not get enough oxygen. The brain cells are then damaged or destroyed.
- Stroke victims will need to follow the guidelines for a diet that is the same as those people with CHD.

Obesity

- Obese people have a much higher risk of developing health problems including increased risk of coronary heart disease, high blood pressure, Type 2 diabetes, joint and mobility problems, stroke and some forms of cancer.
- People who are obese often experience shortness of breath when walking or doing physical activity and may have low self-esteem and depression.
- A BMI of 25 to 29.9 means you are overweight. A BMI of 30 to 39.9 means you are obese. A BMI of 40 and above means you are severely obese.
- Those who are obese should:
  - ○ cut down on fatty and sugary foods that provide energy but little nutrients (for example, cakes, biscuits, crisps, sweet fizzy drinks)
  - ○ eat plenty of fruit and vegetables and fibre-rich foods to fill up (for example, wholegrain cereals, wholemeal bread, brown rice and pasta)
  - ○ grill, steam and bake food with little added fat instead of frying
  - ○ use lower-fat versions of foods, such as cheese, milk, spreads and sauces
  - ○ check the labels on food to make sure they have no added sugars, or use reduced-sugar versions of foods
  - ○ increase physical activities – walk up stairs instead of using a lift; go swimming or to the gym.
- 58% of adults in Wales are classified as overweight or obese.
- Rates of childhood obesity in Wales are the highest in the UK, with 35% of children under 16 being classified as overweight or obese.

Anaemia and iron deficiency

If someone does not eat enough iron in their diet, they will become iron deficient or anaemic. This is because they are not producing sufficient haemoglobin in their red blood cells to carry oxygen around their body.

- Symptoms of anaemia include tiredness and lack of energy, shortness of breath and a pale complexion.
- People at risk of developing anaemia are pregnant women, teenage girls with heavy periods, vegetarians and vegans.

Calcium deficiency and bone health

- Calcium is needed for healthy bones and teeth.
- If you do not have enough calcium in your diet you may develop **osteoporosis** and **hypocalcaemia**.

- Long-term calcium deficiency can cause memory loss, muscle spasms, numbness and tingling in the hands and feet, depression and hallucinations.
- To avoid calcium deficiency you should eat calcium-rich foods (for example, dairy products and oily fish) or take a calcium supplement.

Dental caries

Thirty-three per cent of adults in the UK have some dental decay. This can be prevented by:
- reducing intake of sweet and sugary foods
- brushing teeth well after each meal.
- avoiding sweet and sugary drinks, and watering down fruit juice.

In 2013 it was suggested that 41% of 5-year-olds, 52% of 12-year-olds, and 63% of 15-year-olds in Wales experienced dental decay.

> **Exam tip**
>
> Questions on specific dietary needs or nutritional deficiencies could ask you to identify particular foods that someone with a dietary condition needs to avoid and why they need to do this. Make sure that you fully explain what the forbidden foods will do to a person with that particular condition. An example would be if a question asked for information on coeliac disease: you would explain the intolerance of gluten, describe what happens in the intestine, list foods where gluten is found and offer suggestions for alternatives to these products.

Allergies and intolerances

Nut allergy

Allergic reactions can be minor, and often happen within a few minutes of eating the food.
Symptoms:
- Skin rash.
- Itchy eyes
- Runny nose
- Swollen lips, eyelids and face
- Wheezing or coughing.

An extreme reaction is when the throat starts to swell and the person cannot breathe. This is known as **anaphylactic shock**. Someone with a known allergy to nuts will often carry a special pen called an Epipen, which will give them an injection of adrenaline to reduce the swelling.

Nut allergies can be life-threatening.

Lactose intolerance

- A person who is **lactose intolerant** is allergic to the sugar in milk, which is called lactose. They are unable to digest this particular sugar because they lack the correct enzyme in their small intestine.
- Symptoms include bloating, wind, diarrhoea and nausea.
- The condition is not life-threatening but is very uncomfortable for the sufferer.
- People who have this condition will have to avoid dairy products.

> **Exam tip**
>
> An exam question could focus on lactose intolerance, and ask for suggestions to change a recipe to provide an alternative for someone with lactose intolerance. Remember to change all dairy foods in the recipe to specially made lacto-free products, or alternative ingredients such as soya, almond and rice milks.

Now test yourself

TESTED

1 Explain what Type 2 diabetes is, and list two possible long-term health problems that can result from this condition. [4 marks]
2 Describe three symptoms of anaemia, and identify two groups of people at risk of developing this condition. [5 marks]
3 Suggest ways of preventing dental caries in young children. [4 marks]
4 Using the table below, identify foods in the following recipe that are unsuitable for someone suffering from lactose intolerance, and suggest alternatives. [4 marks]

Table 4.2 **Cauliflower Cheese**

Ingredients	Unsuitable ingredients for a lactose intolerant person	Suggested alternative ingredient
1 cauliflower		
25 g butter		
25 g flour		
250 ml milk		
100 g cheddar cheese		
25 g breadcrumbs (for crunchy topping)		
25 g parmesan cheese (for crunchy topping)		

Planning a balanced diet for people with high energy needs

Sports people and athletes

REVISED

Sports people and athletes will need to have a diet that will provide an increase in energy to be able to compete well and maintain their body weight. The basic athlete's diet plan has an energy intake divided into:

- 55% of energy from carbohydrate
- 12 to 15% of energy from protein
- Less than 30% of energy from fat.

They should base their diet on:

- Wholegrain carbohydrates forming the basis of meals, with extra carbohydrates being consumed per day depending on the level of exercise they do.
- An increase in the amount of protein they eat to help post-exercise recovery and repair, as well as building extra muscle.

Figure 4.7 Sports people have high energy needs

Now test yourself

TESTED

1 Plan a meal for a rugby player to eat before a match. Identify the nutrients found in the meal. [4 marks]

5 Calculate energy and nutritional values of recipes, meals and diets

Calculating energy and nutrients

Energy is measured in **Kilojoules (kJ)** or **Kilocalories (kcal)**.

These are the kJ and kcal values of the three macronutrients.

Table 5.1 kJ and kcal values of macronutrients

Source of energy	Energy value in kJ	Energy value in kcal
1 g of pure carbohydrates	15.7	3.75
1 g of pure fat	37.8	9.0
1 g of pure protein	16.8	4.0

To calculate the energy in a particular food, multiply the number of grams of that food by the energy value in kcal for each nutrient that is in the food. You will have to use an online calculator to find out how much of each nutrient is in that particular food.

For example, to work out the amount of kcal in 100 g of cheddar cheese:
- 100 g of cheddar cheese contains 25.5 g of protein, 35 g of fat and 0.1 g of carbohydrates.
- Total energy from protein = 25.5 × 4 = 102 kcal
- Total energy from fat = 35 × 9 = 315 kcal
- Total energy from carbohydrates = 0.1 × 3.75 = 0.375 kcal
- Total energy from 100 g cheddar cheese = 417.375 kcal

If there is more or less than 100 g of cheese in a recipe, multiply the amounts up.

It is possible to work out the amount of energy for each ingredient in a recipe, a meal and a diet in this way.

> **Exam tip**
>
> You may be given a recipe with the total number of grams of protein, fat and carbohydrate in each ingredient. You will need to multiply these by the number of kcal per gram of each macronutrient and then add them up. If the question asks the number of kcals per portion, divide your answer by the number of servings for the recipe.

Calculating energy values in a recipe   REVISED

To calculate the number of kcal in a recipe, take the quantity of each ingredient and find the number of kcals in each one. Then add up the amount of kcals in the whole recipe.

For example, the amount of kcal in a recipe for cauliflower cheese is as follows:
- 250 g cauliflower = 70 kcal
- 100 g cheddar cheese = 417 kcal
- 25 g flour = 85 kcal
- 25 g margarine = 185 kcal
- 250 ml semi-skimmed milk = 117 kcal
- Total = 872 kcal
- Divide by four for 4 portions = 218 kcal per portion.

> **Now test yourself**   TESTED
>
> 1 Calculate the total energy in one scone from the following recipe. [3 marks]
>
> Sultana scones (makes 12)
> - 225 g flour (742 kcal)
> - 50 g margarine (372 kcal)
> - 75 g sugar (295 kcal)
> - 100 g sultanas (275 kcal)
> - 250 ml semi-skimmed milk (115 kcal)

Calculating the energy value of a person's diet   REVISED

To calculate the daily intake of kcals for one person add the total number of kcals consumed in every meal and snack.

Adapting meals and diets

People of different ages, with different lifestyles, medical conditions or diseases, allergies and intolerances all require changes to their diets. People who will require adaptations to their diet are:

- people with different **religious beliefs**, such as Muslims, Hindus and Jews who have dietary laws specifying what they can and cannot eat
- people with different **ethical beliefs**, such as vegetarians and vegans
- women who are **pregnant**
- people trying to **lose weight**
- people with specific illnesses such coeliac disease or CHD
- people with **allergies** and **intolerances** such as a nut allergy or lactose intolerance.

To keep a balanced, healthy diet it is often recommended that fat, sugar and salt are reduced and fibre is increased.

Reducing fat
REVISED

- Choose leaner cuts of meat and check for the fat content of minced beef.
- Grill, bake and steam rather than frying foods.
- Trim excess fat from meat.
- Choose low-fat versions of spreads and dairy foods.
- Reduce the amount of butter and margarine you spread on bread.
- Use alternatives to high-fat mayonnaise for salad dressings.
- Buy canned fish, like tuna and salmon, in brine rather than oils.

Reducing sugar
REVISED

- Reduce the sugar quantity in recipes.
- Use food sweeteners for stewed fruit and hot drinks.
- Use alternative sweet foods such as carrots (carrot cake), ripe bananas, fresh and dried fruits to add sweetness to cakes and biscuits.

Reducing salt
REVISED

- Use herbs, spices and pepper to flavour food.
- Cut down on processed foods with hidden salt, such as bacon, ham, cheese, salted crisps and peanuts.
- Buy reduced-salt versions of foods.
- Use salt alternatives, such as Lo-Salt, to season foods.
- Read labels to identify sodium, sodium bicarbonate, monosodium glutamate and baking powder, which all contain sodium.

Increasing fibre
REVISED

- Eat wholegrain products, such as bread, cereals, pasta and rice.
- Use wholemeal flour, or half wholemeal and half white, when baking.
- Add oats or wheat bran to crumble toppings, pastry and other recipes.
- Add dried fruit to cake recipes.
- Add fresh or dried fruit to breakfast cereals.
- Make smoothies with fresh fruit.
- Eat the skin of fruit and vegetables, for example jacket potatoes.
- Eat porridge for breakfast.
- Add chopped vegetables to pasta sauces or cottage pie.

Exam tip

- One type of question on adapting meals and diets could be simply to list a number of ways that someone could reduce fat, sugar or salt in their diet, or increase fibre. You will need to make a list with a number of answers to correspond with the number of marks for the question.
- Another type may ask you, for example, to adapt a recipe to provide more fibre. You would then have to substitute the existing ingredients for ingredients that are high in fibre.
- Other questions could ask for a change of ingredients for a particular dietary condition such as lactose intolerance or coeliac disease.

Now test yourself

TESTED

1 Suggest three ways someone could cut down on the amount of salt in their diet. [3 marks]
2 Give two reasons why someone may need to cut down on the amount of fat they are eating and suggest four things they could do to change their diet to reduce the amount of fat. [6 marks]
3 Look at the following meal and make three suggestions as to how the amount of fibre in the meal could be increased. [3 marks]

Main Course

Fish baked in a mushroom sauce

Boiled potatoes

Peas

Dessert

Apple crumble and custard

Increasing energy intake

REVISED

If you wish to increase your energy intake, for example if you are a sportsperson training for a specific event like a marathon, you need to increase the amount of energy you are getting by increasing the amount of carbohydrates you are eating.

Sportspeople trying to gain muscle, for example weightlifters, may need an increase in the amount of protein they are eating.

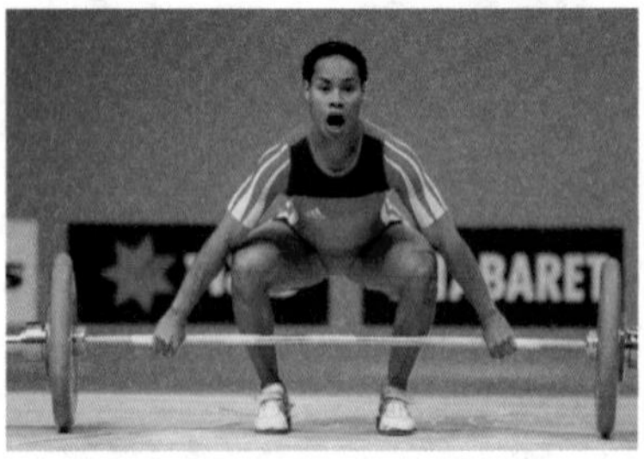

Figure 5.1 A weightlifter will need to increase their amount of protein to gain muscle

Energy balance

Energy balance is when we take in exactly the same amount of energy as we use every day. This will mean that someone doesn't gain or lose weight.

- If you eat too much, and have too much energy from the food you eat, this will be stored as fat in your body and you will gain weight.
- If you use more energy than you eat, you will lose weight.
- When you are young, you are probably more active than when you get older. This means that unless you eat less food as you become less active, you will gain weight.

In order to maintain body weight, energy input must equal energy output.

Now test yourself

TESTED

1 Explain why someone should reduce their energy intake as they get older to avoid gaining weight. [3 marks]
2 Suggest four ways an office worker could adapt their packed lunch to reduce the amount of kcals. [4 marks]
3 Plan a meal for a footballer to eat before a game. Identify which ingredients are going to provide extra energy for the footballer. [6 marks]

Exam tip

Questions on energy intake and output could ask you to design or adapt a day's menu for a specific individual, such as someone trying to lose weight. You would need to look at the meals and suggest alternatives that would provide less kcals. Make sure you discuss the fact that energy intake would need to be less than energy output to lose weight. Also look at the lifestyle of the person, suggesting an increase in exercise or activity to increase their energy output.

6 The effect of cooking on food

Why do we cook food?

- To destroy harmful bacteria
- To make food easier to chew, swallow and digest
- To develop the flavour of foods
- To enable foods to rise, thicken, dissolve and set
- To kill toxins and natural poisons in foods
- To make food look and smell more attractive
- To produce a variety of foods using different cooking methods
- To provide hot food in cold weather

How heat is transferred during the cooking process

- Heat is a type of energy. As heat gives energy to the molecules in food, they start to vibrate and move. The faster they move, the more heat is produced.
- Metal is a good conductor of heat; plastic, glass and cloths are bad conductors of heat.

Heat source

Figure 6.1 Conduction when cooking on a hob

Conduction

This happens when heat is directly touching a piece of equipment or a piece of food, for example frying in a hot pan (see Figure 6.2).

Convection

This only happens in liquids or gas. A **convection current** is caused by the hot liquid rising and allowing colder liquid to drop down. As the liquid rises to the top of the pan, it will begin to cool again, so starts to drop back to the bottom, where it will be heated up again.

Convection currents also happen in ovens. Hot air rises and cooler air falls.

Figure 6.2 Convection currents

Radiation

This occurs through space or air. Radiation transfers energy through space by invisible **electro-magnetic waves**. The waves are either infra-red or microwaves.

- **Infra-red** heat waves are absorbed by the food when they reach it, and they create heat inside the food, which cooks it. This happens when you put food under a grill.

- **Microwaves** are absorbed by the food, making the molecules vibrate and heat up, which then cooks the food. Microwaves pass straight through glass, china and plastic and do not heat them up, so all microwave-safe dishes are made from these materials. Metal will reflect the microwaves and damage the microwave oven, so do not put metal objects into a microwave oven.

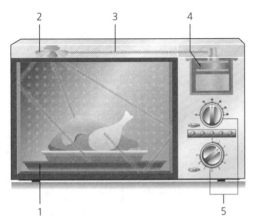

Key

1. Turntable

2. Metal fan to disperse the microwaves

3. Microwaves being created in the magnetron and being pumped into the oven cavity

4. The magnetron

5. Switches for timing and control of microwaves for different functions of the oven

Figure 6.3 How a microwave oven generates radiation

Dishes that rely on more than one method of heat transfer

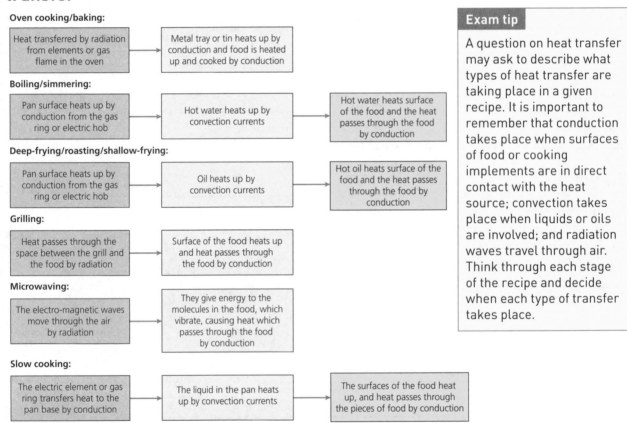

Oven cooking/baking:

| Heat transferred by radiation from elements or gas flame in the oven | → | Metal tray or tin heats up by conduction and food is heated up and cooked by conduction |

Boiling/simmering:

| Pan surface heats up by conduction from the gas ring or electric hob | → | Hot water heats up by convection currents | → | Hot water heats surface of the food and the heat passes through the food by conduction |

Deep-frying/roasting/shallow-frying:

| Pan surface heats up by conduction from the gas ring or electric hob | → | Oil heats up by convection currents | → | Hot oil heats surface of the food and the heat passes through the food by conduction |

Grilling:

| Heat passes through the space between the grill and the food by radiation | → | Surface of the food heats up and heat passes through the food by conduction |

Microwaving:

| The electro-magnetic waves move through the air by radiation | → | They give energy to the molecules in the food, which vibrate, causing heat which passes through the food by conduction |

Slow cooking:

| The electric element or gas ring transfers heat to the pan base by conduction | → | The liquid in the pan heats up by convection currents | → | The surfaces of the food heat up, and heat passes through the pieces of food by conduction |

Figure 6.4 Dishes that rely on more than one method of heat transfer

Exam tip

A question on heat transfer may ask to describe what types of heat transfer are taking place in a given recipe. It is important to remember that conduction takes place when surfaces of food or cooking implements are in direct contact with the heat source; convection takes place when liquids or oils are involved; and radiation waves travel through air. Think through each stage of the recipe and decide when each type of transfer takes place.

Now test yourself

TESTED

1 Explain why handles on saucepans are not usually made of metal. [2 marks]
2 Give an example of a form of cooking that shows two different methods of heat transfer, and explain how the heat is transferred in each of these methods. [4 marks]
3 Identify a method of cooking that shows the use of convection currents to transfer heat. Describe how a convection current is created, using diagrams. [3 marks]

Selecting appropriate cooking methods to conserve or modify nutritive value and improve palatability

Conserving nutritive value

REVISED

Table 6.2 How nutrients are destroyed

Nutrient	How easily is it destroyed?
Protein	Not destroyed by heat, but chemical changes result in denaturation
Carbohydrates	Not destroyed by heat, but chemical changes result in starch degradation
Fat	Not destroyed by heat, but some methods of cooking add fat (e.g. frying) and some reduce fat content (e.g. grilling)
Fat-soluble vitamins A, D, E and K	Will leach out of the food into the fat if foods containing these vitamins are cooked using fat
Water-soluble vitamins B and C	Will be damaged by heat and will dissolve into the water the food is cooked in
Minerals	Not affected by cooking processes

The main foods that could lose nutrients during cooking are those containing vitamins. Many vitamins are found in fruit and vegetables, so these must be cooked carefully to ensure as many vitamins are kept in the food as possible.

Ways to cook foods containing Vitamin A to conserve nutritive value

- Serve the fat as part of the dish, for example a stir-fry uses a minimum amount of fat and the food is coated with the fat as it is cooked, so is usually eaten as part of the dish.
- Steam or boil vegetables
- Use the oil or fat that has been used to cook foods to make gravy or a sauce.

Ways to cook foods containing water-soluble vitamins to conserve nutritive value

These are easily destroyed by heat, and dissolve in water during cooking. Vitamin C is also destroyed when it is exposed to oxygen.

- Buy the foods as fresh as possible, as these will have the most Vitamin C.
- Prepare them at the last possible minute to reduce the amount of time they are exposed to the air.
- Select a cooking method that uses as little water as possible; steaming rather than boiling, microwaving, or frying in a little oil.
- Cook them for as short a time as possible and serve immediately as the longer you keep them hot, the more Vitamin C will be destroyed.
- Use the cooking liquid to make a sauce or gravy or make a soup from the liquid.
- Bake root vegetables such as potatoes or sweet potatoes whole to retain their Vitamin C.

Figure 6.5 The way in which you cook will affect nutritive value

Modifying nutritive value

- **Reduce fat, sugar or salt:** modify the recipe by using alternative ingredients.
- **Increase fibre content:** add or substitute wholemeal ingredients, fresh fruit or vegetables such as adding grated vegetables, (e.g. carrots) into a meat dish like lasagne, to give vitamins and fibre.
- **Add extra nutrients:** add an egg or chopped meat, such as bacon, to a macaroni cheese dish for extra protein or add powdered products such as wheat germ to stews and soup for extra protein.

Improving the palatability of a dish

There are times when food palatability needs to be improved.
- When elderly people have small appetites, make the food as tasty as possible to encourage them to eat. This will increase saliva flow in the mouth, so the food is digested more easily.
- Tenderise meats to make them easier to chew.
- People undergoing some cancer treatments, such as chemotherapy, may experience problems with their mouths, and food may not taste very nice. Slight sweeteners can help improve the taste of food, making it more palatable.
- Adding fat to foods: most people enjoy the mouthfeel (the creamy satisfying feel) of fatty foods, so tend to eat more and enjoy them more.

Exam tip

Questions on this section are more likely to concentrate on the preservation of nutrients. A possible question could ask how you can preserve as much Vitamin C as possible when preparing and cooking cabbage. Remember to include information about the buying, storing and preparation of the cabbage as well as suggesting ways of cooking such as steaming, stir-frying or microwaving, to conserve the Vitamin C.

Now test yourself

1 Identify two vitamins that are destroyed by cooking in fat. [2 marks]
2 Explain how you would prepare and cook spinach to preserve as much Vitamin C as possible. [4 marks]
3 If you are preparing a meal for someone who has difficulty chewing, identify three things you could do to the food to allow them to enjoy the meal and be able to eat the food easily. [3 marks]
4 Discuss why it is important that extra protein could be added to a meal for someone who has been ill and needs to build up strength. Suggest at least three ways this could be done. [6 marks]

The positive use of micro-organisms

Some micro-organisms are useful during food production.
- Yoghurt is made using friendly bacteria, which feed on the lactose (sugar) in the milk and convert it into lactic acid, which denatures the milk proteins and thickens the product. The acid gives yoghurt its sharp taste.
- An enzyme called rennet is added in cheese making, which separates the milk into curds and whey. The curd – the thick, white part – is used to form the cheese. Blue cheeses, such as Stilton, have moulds added to them to create the veins of blue that are found in the cheeses.

Figure 6.6 Yoghurt is made using friendly bacteria

- Meat products such as salami, chorizo and French saucisson are made from fermented meat. Bacteria are used to change the acidity of the meat and prevent harmful bacteria spoiling it. This results in denaturation of the protein.
- Alcoholic drinks use yeast and the resulting sugar fermentation to produce the alcohol.
- Yeast is used to make bread rise.
- Quorn is a meat substitute made from a mycoprotein, which is derived from a fungus. A specific vegan Quorn is now available that does not contain egg albumin.

Figure 6.7 Stilton cheese has mould added to create the blue veins

Exam tip

There may possibly be a question asking you to name some products where micro-organisms are used to create foods for human consumption. Remembering three different foods where this is shown will almost certainly be sufficient to answer a question on this topic.

Now test yourself   TESTED

1 Name three foods that use micro-organisms to produce the final result.   [3 marks]
2 Explain how bacteria are used to make yoghurt.   [4 marks]

Working characteristics, functional and chemical properties of ingredients

Proteins   REVISED

- Eggs are an HBV protein food source and exhibit all protein characteristics.
  - **Coagulation** when heated: this is shown when eggs are cooked and change structure.
  - **Foam formation** (or **aeration**): this is shown when eggs are whisked or beaten to incorporate air.
- **Gluten formation:** this is shown when bread dough is worked and kneaded to stretch the gluten strands in the flour and create a structure for the bread.
- **Denaturation:** this is when the protein unravels and a new structure is formed. Denaturation is caused by heat, **agitation** such as whisking or tenderising with a meat mallet, and by the effect of acid, for example when a marinade is used.

Figure 6.8 Products showing coagulation, foam formation and denaturation of egg protein

Table 6.3 Functions of eggs

The function of eggs	Why the egg is used	Examples of recipes using this function
To bind ingredients together	The protein in the egg will coagulate (become thick and set) when heated, so it holds the other ingredients together.	Fish cakes Potato croquettes
To trap air	The protein in the egg will stretch when it is whisked (denaturation) and this will allow air to be trapped as little bubbles (foam formation or aeration). When the product is cooked it will solidify around the air.	Meringues Whisked sponges, such as Swiss rolls Mousses and soufflés
To thicken products	As the protein in the egg is heated it coagulates and thickens the product.	Egg custard Quiche Lorraine
To coat products	The product is dipped in egg and breadcrumbs before it is fried. The egg coagulates and seals the product in the crispy coating.	Scotch eggs Fish in breadcrumbs
To create an emulsion	The protein in the egg will stop oil and water separating. This is called an emulsion.	Mayonnaise, when the oil and vinegar are held together by the egg yolk
To glaze products	Egg is brushed on the surface of a product, and when it is cooked the egg will form a shiny surface (coagulation).	Pastry toppings on pies Bread rolls or a loaf of bread

Exam tip

A question could ask you to explain what happens when you cook an egg. You will need to explain protein denaturation and write about what happens using the correct scientific wording, for example, 'When the egg is cooked, heat causes the DNA to unravel and create a new structure. This means that the colour and texture of the egg white changes, and the egg yolk will set. This is a reaction that cannot be reversed.'

Figure 6.9 Raw egg and cooked egg showing the effects of protein denaturation

Now test yourself

TESTED ☐

1 Give an example of a recipe where you might put meat in a marinade prior to cooking to improve the tenderness and flavour of the meat. Explain how the marinade makes the meat more tender. [4 marks]

2 Identify two recipes where air is trapped due to the whisking of eggs. [2 marks]

3 Explain how kneading bread dough improves the structure of the final cooked product. [4 marks]

Carbohydrates

Starches

- **Gelatinisation:** when mixed with liquid and heated, starch granules swell and burst, releasing the starch, which then forms a gel and thickens a product.
- **Dextrinisation:** starch converts to sugar when it is heated. This is seen when bread is toasted and turns brown.

Table 6.4 Uses of different flours

Type of wheat flour	Reason for use	Examples of recipes using this flour
Self-raising flour	This is a soft flour that contains baking powder as a raising agent. It contains less than 10% protein (gluten), so is not stretchy.	Cakes, biscuits and scones
Soft plain flour	This contains less than 10% gluten, so produces a soft dough that is not stretchy.	Shortcrust pastry Biscuits
Strong plain flour	This contains more than 10% of the protein gluten, so will produce a very stretchy dough when it is worked or kneaded, as the gluten will form stretchy strands to trap the carbon dioxide produced by the yeast.	Bread and bread products
Durum wheat flour	This is a special flour used to make pasta. It contains tough gluten, so will not stretch very well.	Fresh pasta

Sugars

- Add sweetness to a product.
- Add texture to a product. Brown sugar has larger crystals to give a crunch. Sugar softens gluten making a softer cake product.
- Add colour to a product: sugar **caramelises** when heated and turns brown.
- Trap air: when sugar is creamed with fat, air is trapped in the mixture.

Exam tip

Questions on the functions of carbohydrates could ask for a detailed description of how a roux sauce thickens. Make sure you include all details of how the starch granules begin to cook when mixed with the fat and continue when the liquid is added and heated to boiling point, which is when the granules burst and release the starch, absorbing the liquid and thickening the sauce. Don't forget to include the word gelatinisation in your description as the correct term for this.

Now test yourself

1 Identify two types of flour that are easily available to buy and state which recipes you would use these flours for, explaining why they are suitable. [4 marks]
2 List three functions of sugar, giving examples of a recipe where each of the chosen functions is shown. [6 marks]

Fats and oils

- **Shortening:** gives a product a crumbly texture, such as in shortcrust pastry or biscuits. The fat coats the flour particles with a waterproof layer. This stops the gluten forming long strands and makes the final baked product have a soft texture because the fat has shortened the gluten strands.
- **Aeration:** when fat is creamed with caster sugar it will trap air. This will form a stable foam, which is then cooked. The trapped air will rise and be trapped by the gluten in the flour as the cake is cooked.
- **Plasticity:** different fats melt at different temperatures. Some products are created with lower melting points so that they can spread straight from the fridge. Other fats, such as butter, have a higher melting point, so will be solid when taken from the fridge, but will soften if left in a warm room.
- **Emulsification:** fats are **hydrophobic**. This means that they will not mix with water, but tend to form large globules when mixed with water or a liquid.

Chemical reactions in fruit and vegetables

- **Oxidisation:** when fruits and vegetables are cut or peeled, the surface of the fruit or vegetable is exposed to the air. The oxygen in the air reacts with the fruit or vegetable.
- **Enzymic browning:** when the cells in the fruit or vegetable are cut, they release enzymes that react with oxygen, turning the fruit or vegetable brown.

> **Exam tip**
>
> A possible question on fats and oils could ask for the definition of plasticity. A full explanation is needed, such as, 'Plasticity is the ability of fats to melt at different temperatures. Each type of fat has its own melting point, which makes it unique. This means that different fats are suitable for different recipes. Some fats are deliberately created to be softer at room temperatures, so they can spread on bread easily. An example of this is "I Can't Believe it's Not Butter". Fats such as real butter are solid in the fridge, but soften at room temperature.' This shows that you really understand the term plasticity and will gain full marks.

Figure 6.10 A cut apple showing enzymic browning.

Now test yourself

1 Explain how fat can help to aerate a cake mixture. [3 marks]
2 Why would adding margarine to a recipe increase the nutritional value of the final outcome? [2 marks]
3 Describe the chemical changes that are taking place when an apple is cut and turns brown if left in the open air. [3 marks]

Why some recipes do not succeed and how to remedy situations

Following a recipe exactly should ensure a successful outcome. Unsuccessful outcomes can be because of:

- Using the wrong ingredients: for example, plain flour instead of self-raising flour.
- Using out-of-date ingredients: for example, dried yeast that is past its sell-by date.
- Incorrect weighing of ingredients: for example, adding too much sugar, not enough flour or too much liquid.
- Incorrect mixing of ingredients: for example, not working a dough enough so a bread product has an uneven rise.
- Overworking of the product prior to cooking: for example, overworking pastry causing an oily dough.
- Not allowing sufficient resting or rising time: for example, not allowing bread dough to rise, so the final product is flat and heavy-textured.
- Incorrect oven temperature: for example, not allowing the oven to pre-heat, or having the oven too hot or too cold, resulting in an over-or undercooked outcome.
- Insufficient stirring during heating, resulting in a burnt or lumpy sauce.

Sometimes it is possible to remedy a situation, but often it is not. Knowing where you went wrong will help you to understand how to prevent the same fault happening next time you make that recipe.

Figure 6.11 Incorrect mixing of ingredients can cause a cake to rise unevenly

Now test yourself

TESTED ☐

1 Give two reasons a cake may have sunk in the middle, and explain why these reasons caused the cake to sink. [4 marks]
2 Describe what could have happened to a pastry case that has shrunk when it was cooked. [2 marks]
3 How can I ensure that a loaf of bread does not have a dense, uneven texture when cooked? [4 marks]

Exam tip

A question could be specific to a particular recipe, for example it could ask you to give reasons a loaf of bread has not risen. Look at the number of marks available, and decide how many reasons you need to give. Remember, you will get one mark for the reason and one mark for the explanation of how this has prevented the bread rising. A four-mark question would need two reasons, with a full explanation of each reason.

7 Food spoilage

Storing food correctly

In order to minimise food spoilage, food waste and contamination of foods that may cause illness, it is important to store foods correctly.

Where food is stored depends on the type of food.

In the fridge

Foods that should be stored in the fridge include:
- fresh foods, such as milk, cheese, eggs, meat, fish and ready meals
- salad foods, some fruit and vegetables
- foods that have 'keep refrigerated' on the label
- some foods that need refrigerating only after they have been opened (check the label).

Rules for storing foods in the fridge
- The temperature of the fridge should be between 0 °C and 5 °C.
- Never put hot food in the fridge, as it will raise the temperature. Allow it to cool down first.
- Do not overload your fridge, as cool air must be able to circulate.
- Throw foods away that are past their use-by date.
- Always store raw meats and fish on the bottom shelf to avoid cross-contamination from drips from the raw food.
- Cooked meats should be on the top shelf, away from raw meats.
- Keep foods wrapped and covered, or in airtight containers to avoid cross-contamination.
- Clean the fridge regularly.
- Do not open the door too often, as this will raise the temperature.

In the freezer

- Freezing foods slows down bacterial growth by freezing the water in the food.
- When the food is defrosted the water defrosts and the food may begin to break down.
- Bacterial activity will begin when the food enters the **danger zone** of between 5 °C and 63 °C, so defrosted food must be treated the same as fresh food, and not refrozen.

Freezers have star ratings that show how long food can be frozen for.

Table 7.1 Freezer star ratings

Star rating	Temperature	Storage time
One star (*) (an ice box in a fridge)	–6 °C	Up to one week
Two star (**)	–12 °C	Up to one month
Three star (***)	–18 °C	3–12 months

Rules for storing food in the freezer

- Food must be well wrapped in freezer bags or suitable airtight containers, to avoid freezer burn, which happens when air meets frozen food.
- Label and date all food so you know when to use it by.
- Overloading the freezer will affect the temperature, and may cause foods to spoil.
- Check the temperature regularly with a freezer thermometer.
- Put newly frozen foods at the back or bottom of the freezer so the older foods are used first. This is the FIFO rule (First in First out).
- Follow recipes for instructions on freezing.

Dry storage of foods

REVISED

- Packaged canned or bottled foods need to be stored carefully to avoid contamination and maintain the best quality.
- Store fresh foods, such as root vegetables like potatoes and carrots, in a dry, cool, well-ventilated place.
- The temperature of the room should be between 10 °C and 20 °C, but the cooler the room is the better.
- The food should be off the floor to prevent pests and animals contaminating the product.
- Keep out of direct sunlight.
- Store food in its in their original packaging, or in airtight containers.
- All foods should be dated so they are used within the recommended time.
- Place newest foods behind the oldest to ensure they are used in date order.

> **Exam tip**
>
> Questions on this section of the course may be included within a larger question. For example, a question could ask you to describe how you would keep certain foods safe when preparing, cooking and serving a buffet. You would need to think about the storage of the food both before preparation and after the food has been prepared. You would include details of the temperature of the fridge/freezer, where you would place the food in the fridge/freezer, how it would be wrapped, that you will check date labels and how you would avoid contamination.

Now test yourself

TESTED

1 State the correct temperature for a freezer. [1 mark]
2 Explain why raw meat should be stored at the bottom of the fridge. [3 marks]
3 Describe four conditions needed to store dried goods, such as flour. [4 marks]
4 List two reasons why it is important to place newer foods behind older foods in the storeroom. [2 marks]

Date marks and labelling

All purchased foods will be marked with a date mark.

- **Use-by:** food must be consumed by this date. Food should be thrown away if it has passed its use-by date.
- **Best-before:** food will be at its best quality before this date. The food can still be consumed safely after this date, but it may not taste at its best.
- **Sell-by:** this is for the supermarket or shop to use to ensure food is sold at its best quality and freshness.

Exam tip

Make sure you can clearly explain the difference between 'use-by' and 'best-before' dates. Be able to explain why food should not be consumed after its use-by date, as it could cause food poisoning and illness, but that it can be consumed after its best-before date, but the food may not be of such high quality.

Now test yourself

TESTED

1 Explain what the FIFO (First in First out) rule is and why it is so important to follow this rule.   [3 marks]
2 Give three reasons it is important to clearly label foods before they are put in a freezer.   [4 marks]

Food spoilage

Enzyme action, mould growth and yeast production

REVISED

Food spoilage is caused by **bacteria**, **yeasts**, **moulds** and **enzyme action**. These can make food unsafe to eat by being in the food, or by producing waste products, toxins or poisons that contaminate the food.

In order to prevent growth of micro-organisms, you need to remove one or more than one of the conditions needed for them to grow, which are:

- warmth
- moisture
- time.

Figure 7.1 Mould on a loaf of bread

Table 7.2 Food spoilage

Type of spoilage	Description	How does it work?	Ways of prevention/control methods
Enzymic action	Enzymes are chemicals found in both plants and animals. They are made of proteins. They can then cause unwanted changes to foods, such as a colour change. Enzymes will also ripen or over-ripen fruit by turning the starch into sugar.	When the foods are killed or harvested, the enzymes are activated.	• Blanch fruit and vegetables in boiling water for one minute to inactivate the enzymes and then plunge them into cold water to stop the cooking process. • Place the fruit, for example an apple, into acidic conditions, such as lemon juice, to prevent the fruit browning. • Cover the fruit or vegetables with cling film to reduce contact with the air.

Type of spoilage	Description	How does it work?	Ways of prevention/control methods
Mould growth	Moulds are a type of fungi. They can be blue, green, black or white.	They reproduce by making spores, which are airborne and land on food. If the conditions are correct they will grow. Moulds can grow on foods that are slightly acid or alkaline, sweet, salty, dry or moist.	• Keep food chilled. • If the food is not refrigerated, store in a cool, dry place. • Store in acidic conditions, for example in vinegar, if suitable. • Heat food to above 100 °C to prevent heat resistant spores being produced.
Yeast production	Yeast is a type of fungi found in the air, in soil and on the skin of certain fruits.	It reproduces by multiplying one cell into two. This is called budding.	• Keep food chilled (reduce warmth): yeasts are dormant or inactive at cold temperatures. • Keep dried or fresh yeast for bread making away from moisture as this will activate the yeast. • Yeasts are killed at high temperatures.
Bacteria	Bacteria are extremely small organisms that can only been seen under a microscope. They are everywhere: in water, in the air, on humans and animals.		• Store food in the fridge between 0 °C and 5 °C. • Chill cooked food rapidly, in less than 90 minutes, if it is to be stored for later consumption, and store in the fridge or freezer. • Leftover food should be eaten within 24 hours and only reheated once, as reheating will activate the bacteria again. • Store foods in acid, for example vinegar, or a high amount of salt or sugar.

Exam tip

Remember that taking away any of the conditions that yeast, mould and bacteria need to grow will reduce the contamination of the food, but will not completely prevent it. For example, chilling and freezing foods makes the organisms dormant, but when the temperature rises again they will begin to grow. When answering a question on prevention of contamination by these organisms make sure you include details of how to remove any of the conditions needed for these to grow, and why it will slow them down.

Now test yourself

TESTED ☐

1 Explain how enzymes alter the appearance of a cut apple. [2 marks]
2 Describe two ways that enzymic browning can be controlled. [4 marks]
3 Identify three conditions necessary for bacteria to grow. [3 marks]
4 Why is 37 °C an optimum growth temperature for food-poisoning bacteria? [3 marks]

Signs of food spoilage

- Spoiled food is also referred to as food that has 'gone off'.
- Fresh food begins deteriorating as soon as it is picked or slaughtered.
- Food can be spoiled by moisture loss over time as well as the action of micro-organisms.
- Spoiled food will begin to show changes to the:
  - texture: may become slimy
  - flavour: may become sour
  - smell: may smell sour or bad
  - appearance: may look over-dry or over-wet, or look wrinkled or discoloured, or mould may have grown on the food.
- Food contaminated by bacteria may not show any changes in texture, flavour, smell or appearance.
- Protein foods will rot and smell bad.
- Fats and oils go rancid and will begin to smell and have a bad taste.
- Carbohydrate foods will become slimy, and smell and taste unpleasant.
- Fruit and vegetables will grow mould and begin to rot.
- Food spoilage can also be caused by poor handling, cross-contamination by the handler, or by incorrectly storing the food.

Exam tip

When answering a question on signs of food spoilage, think about the three micro-organisms that contaminate food – yeast, mould and bacteria – and remember that only two of them will change the appearance, smell and taste of the food. Think about rotten fruit, cheese or bread with mould on or bad-smelling fish or meat, and describe what you can see or smell.

Now test yourself

1 Identify three ways you could tell a food was spoiled. [3 marks]
2 Explain what the action of enzymes will do to fruit that is left at room temperature for a week. [4 marks]

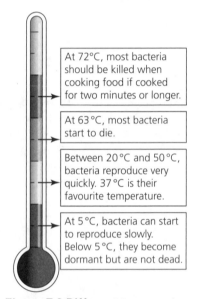

Figure 7.2 Different temperatures and bacteria growth

At 72°C, most bacteria should be killed when cooking food if cooked for two minutes or longer.

At 63°C, most bacteria start to die.

Between 20°C and 50°C, bacteria reproduce very quickly. 37°C is their favourite temperature.

At 5°C, bacteria can start to reproduce slowly. Below 5°C, they become dormant but are not dead.

The role of temperature, time, pH and moisture in the control of bacteria

Temperature

- The optimum temperature for bacterial growth is 37 °C.
- They are most active in the **danger zone**, which is between 5 °C and 63 °C.
- Bacteria will multiply at different rates, depending on the temperature.

Time

- Bacteria multiply by dividing into two.
- This can happen as often as every 10–20 minutes.
- One bacterial cell can multiply to millions after 24 hours.
- Incorrect storage or allowing food to be left in warm, moist conditions can allow food poisoning bacteria to multiply.

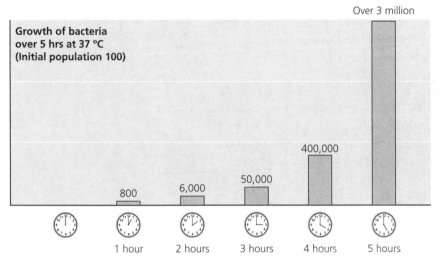

Growth of bacteria over 5 hrs at 37 °C (Initial population 100)

Over 3 million

400,000

50,000

6,000

800

1 hour 2 hours 3 hours 4 hours 5 hours

Figure 7.3 Growth of bacteria at 37 °C

pH

- The pH scale shows how acid or alkaline a substance is.
- Below pH 7 indicates acidity. pH 7 is neutral. Above pH 7 shows alkalinity.
- Bacteria prefer neutral conditions, so are unable to grow in acid or alkaline conditions.

Moisture

- Bacteria need moisture to grow.
- Foods such as soups, gravy and sauces are therefore **high-risk foods**, depending on the ingredients used to make them.

Exam tip

To gain full marks in a question on the control of bacteria in food production and storage you need to be able to explain the conditions needed for bacterial growth. Include reference to the optimum temperature being 37 °C, which is the human body temperature. Remember that foods contaminated with bacteria will not look, smell or taste bad, so the only way to prevent contamination is by exercising proper control over foods to ensure they do not become contaminated.

Now test yourself

1 Explain how bacteria multiply. [4 marks]
2 Why is it important to store foods below 5 °C, or heat them to over 63 °C? [2 marks]
3 Give two examples of moisture-rich, high-risk foods. [2 marks]

Bacterial cross-contamination

Types of bacterial cross-contamination

REVISED

- **Cross-contamination** is when bacteria are transferred from one source, for example raw food, to another cooked source, which may lead to food poisoning.
- Cross-contamination can occur in several ways:
  - from high-risk raw foods to cooked food by using the same chopping board or knife without washing the board or knife in between
  - by storing raw meat or fish above cooked foods in the fridge, so drips from the raw foods fall onto cooked food
  - by placing incorrectly covered raw food next to cooked food in the fridge
  - incorrect or infrequent hand washing during food preparation or after sneezing or going to the toilet, resulting in transfer of bacteria
  - dirt or soil from vegetables being transferred to other foods due to insufficient washing of vegetables or equipment
  - continuing to prepare foods while you are suffering from a stomach upset, as your hands can have bacteria on them, which will be transferred to the food.

How to prevent bacterial cross-contamination

- Wash chopping boards and equipment with hot, soapy water after preparing raw meats.
- Use colour-coded chopping boards, usually red, for raw meat preparation.
- Use clean equipment for each stage of food preparation.
- Thoroughly wash your hands after preparing raw meat, raw fish or dirty vegetables and after going to the toilet or blowing your nose.
- Store raw meat and fish in a covered container on the bottom shelf of the fridge, away from cooked foods.
- When cooking raw meat, put all meat in the pan at the same time and do not add more during the cooking process. Use a separate pan for any extra raw meat.
- Use clean cloths to wipe surfaces and equipment.

> **Exam tip**
>
> A question on the prevention of cross-contamination could possibly be included in a more detailed question about food preparation. You need to be able to explain how cross-contamination occurs and how you would prevent it. An example would be to use colour-coded chopping boards, as raw meat can cross-contaminate cooked foods.

Now test yourself

TESTED

1 Give an example of when cross-contamination can occur during food preparation, and explain what you would do to prevent your example occurring. [4 marks]
2 Why is it important that someone suffering from a stomach upset does not prepare food until they are fully recovered? [3 marks]
3 Discuss how you would store raw fish in the fridge, and list two reasons it is so important to follow this rule. [4 marks]

Now test yourself answers and quick quizzes at **www.hoddereducation.co.uk/myrevisionnotes**

Preservation

Food is preserved to stop it deteriorating and going off and to extend its shelf life.

Freezing

- Cold temperatures will not kill micro-organisms, but will slow them down or make them dormant or inactive.
- The water in the food will become frozen, so the micro-organism has no access to moisture, which is needed for growth.
- As food defrosts, moisture becomes available, and the micro-organisms will begin to grow.
- Never refreeze defrosted or thawed food as it will contain harmful micro-organisms.
- Freeze when at their best quality.
- Blanch fruit and vegetables before freezing to kill harmful bacteria on the surface.
- Wrap food in freezer bags, label and date before placing in the freezer. Incorrectly wrapped foods may get freezer burn.

Figure 7.4 Foods containing lots of water do not freeze well

Pickling

- Pickling involves covering the food in pickling vinegar and changing the pH levels to below pH 7, creating an environment where micro-organisms cannot grow.
- Onions and gherkins, fish such as herring, and protein foods such as eggs can be pickled.
- Foods that are to be pickled are often soaked in a salt solution first to remove some of the water from the food.
- Pickling spices are added to the vinegar for extra flavour.
- Foods are often cooked in the spiced vinegar. They are put into sterilised, sealed glass jars and left to mature for several weeks before they are ready to eat. Sometimes sugar is added to sweeten the pickled food.
- Pickling helps reduce enzymic action.

Jam making

- Jam making uses sugar and water, along with fruit. The high sugar content creates an environment where micro-organisms cannot grow.
- The ingredients are boiled, until they reach 105 °C, which is the setting point of jam.
- This also destroys any yeast that may be present, which can spoil jam.
- Jam is placed in sterilised glass jars and sealed immediately to prevent any contamination with other micro-organisms.
- Unopened jars of jam should be stored in a cool, dry cupboard out of direct sunlight.
- Once a jar of jam is opened, it should be stored in the fridge and consumed within one month.

Bottling and canning

- Vegetables or fruits are put into sterilised sealed jars, or sealed cans, with either a salt solution (brine) or a sugar solution.
- These jars or cans are then boiled at high temperatures to kill micro-organisms.
- There is no air in the jars, which also helps to preserve the food.
- The jars are left to cool, and the unopened jars will last for several months.
- Once opened, the jars must be stored in the fridge and used within a few weeks.

Vacuum packing

- Foods are placed in special packs and the oxygen (air) is removed. This means that most micro-organisms cannot grow.
- The food then is chilled.
- Meat, fish, vegetables and fresh pasta can be bought as vacuum-packed products.
- Once opened, vacuum-packed foods must be treated the same way as fresh products.

Figure 7.5 Vacuum-packed meat

> **Exam tip**
>
> Some questions ask for suggestions on how to preserve foods. Remember to include information about the conditions needed for micro-organisms to grow – moisture, warmth, time and food, along with oxygen. Taking away any one of these conditions will stop growth. Then list the ways that this can be done. Explain with each method which condition you are removing, and how the process is carried out.

Now test yourself

1 Explain how fruits and vegetables are bottled or canned and how this process reduces the chances of food becoming spoiled by the action of micro-organisms. [4 marks]
2 List three fruits that can be used to make jam, and explain the process for jam making. Include details of storage. [4 marks]
3 Why is it important to wrap food correctly before placing it in the freezer? [2 marks]
4 Explain why food that has been defrosted cannot be refrozen. [3 marks]

Food poisoning

Food poisoning is when bacteria enter the body through contaminated food or water and start to reproduce, which causes illness either from the bacteria themselves, or from the toxins they produce.

Reasons for food poisoning include:
- poor hygiene during food preparation and handling
- incorrect cleaning procedures
- incorrect temperature control during food storage (chilling) or when serving hot food
- incorrect cooking times, or not allowing the centre of a cooked food to reach the correct temperature of 72 °C for at least two minutes
- reheating food for an insufficient length of time, or reheating food more than once.

Signs of food poisoning

REVISED

Salmonella

Table 7.3 Food poisoning: Salmonella

Symptoms	Nausea, vomiting, fever, diarrhoea, headache and abdominal pain Elderly and very young children can become extremely ill and have died from salmonella poisoning.
Which foods contain this bacteria	Chicken, poultry, some dairy foods and raw or undercooked eggs
Where it is found	Dirty water, raw foods, transferred from people or pests
How to prevent contamination	1 Wash hands before and after handling raw meat. 2 Never wash raw chicken as this will spread bacteria. 3 Store raw meat at the bottom of the fridge. 4 Cook meat and eggs thoroughly. 5 Wash all fruit and vegetables thoroughly before use.

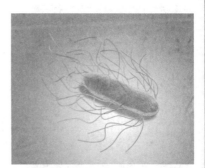

Figure 7.6 Salmonella bacteria

Campylobacter

Table 7.4 Food poisoning: Campylobacter

Symptoms	Stomach cramps, fever, vomiting and bloody diarrhoea
Which foods contain this bacteria	Undercooked or raw meat, particularly poultry, untreated water and unpasteurised milk
Where it is found	Animals, pests, untreated water, sewage
How to prevent contamination	1 Wash hands before and after handling raw meat. 2 Never wash raw chicken as this will spread bacteria. 3 Store raw meat at the bottom of the fridge. 4 Cook meat thoroughly. 5 Keep animals away from foods. 6 Wash all fruit and vegetables thoroughly before use.

E-coli

Table 7.5 Food poisoning: E-coli

Symptoms	Severe abdominal cramps, fever, fatigue, bloody or watery diarrhoea, nausea and vomiting
Which foods contain this bacteria	Raw meat, cooked meat and gravy, or other meat products, unpasteurised cheeses and juices, raw fish and oysters
Where it is found	Raw meat that has become contaminated, waste from animal intestines, dirty water and human waste
How to prevent contamination	1 Wash hands before and after handling raw meat. 2 Never wash raw chicken as this will spread bacteria. 3 Store raw meat at the bottom of the fridge. 4 Cook meat thoroughly. 5 Correctly store cooked meat, cheeses, raw fish and oysters and check packaging and sell-by dates. 6 Never reheat gravy more than once.

Staphylococcus

Table 7.6 Food poisoning: Staphylococcus

Symptoms	Diarrhoea, vomiting, can cause people to collapse
Which foods contain this bacteria	Egg products, milk, cream, cooked meat and meat products, chicken, salads containing chicken, tuna and egg
Where it is found	On people in the nose, on the skin in the mouth, on the hair, from cuts, burns, scratches and skin infections, from raw, untreated milk
How to prevent contamination	1 Wash hands before and after handling raw and cooked meat. 2 Cover or tie back hair, do not sneeze over food and wash hands after blowing your nose. 3 Cover all cuts and scratches with blue plasters. 4 Store cooked foods correctly, cover and place in the fridge.

Exam tip

You will need to know about the different types of food poisoning bacteria, where and how contamination can occur, and how to prevent it. Make sure you can describe the symptoms of each type, and which foods contain these bacteria. List them in your answer. Learn the rules for hygienic ways of working in the kitchen as these will prevent contamination of all the different types of bacteria.

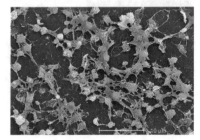

Figure 7.7 Staphylococcus bacteria

Now test yourself

TESTED

1 List two symptoms of E-coli poisoning, and state where this bacteria is found. [4 marks]
2 Give three ways to prevent and reduce the spread and contamination of foods by Staphylococcus bacteria. [3 marks]
3 Explain how bacteria reproduce, and give three conditions that are needed for growth of bacteria. [4 marks]

Food wastage

- The main types of food thrown away include:
  - prepared foods such as pasta and rice
  - fresh vegetables, salads and fruit
  - bread and cakes
  - ready-made meals and take-away foods, often unopened.
- Foods are thrown away because:
  - they are past their 'sell-by' or 'use-by' dates
  - the food smells or looks 'off'
  - the food has gone mouldy
  - too much food was prepared, or was left on a plate during the meal.

Effect on the environment

REVISED

- Waste food placed in rubbish bins will attract pests and bacteria.
- If it is taken to a landfill waste site it will rot down and produce methane gas, which is a greenhouse gas that contributes to global warming and is more harmful than carbon dioxide.
- Many local authorities will provide special bins for recycling food waste. This is then collected and used to create energy in special biomass plants, or used to make compost that is then resold.

Financial implications of waste

REVISED

- Wasting food wastes money.
- Estimates are that food wastage costs each family in the UK up to £470 a year, up to £700 a year for families with children.
- Disposing of food waste costs local councils money that could be better spent elsewhere.

How to reduce food waste

- Plan meals for the week and make a list of foods needed. Only buy those foods when you shop.
- Avoid 'buy one get one free' or 'two for one' offers unless you can store or freeze the second product for use later on.
- Use foods that you already have in the freezer or cupboard instead of buying new foods.
- Only cook enough food for each meal, using correct portion control.
- Freeze leftover foods in portions for another meal.
- Use leftover food to create another dish, for example a meat pasta sauce can be used in a cottage pie; leftover chicken or vegetables can make a soup.
- Use excess fruit to make jam or jellies, or cook to make a pie or crumble, which can be frozen.
- Freeze fresh meat or fish that is near its use-by date for use in meals later on.

> **Exam tip**
>
> A question on food wastage may be an essay-type question on ways to reduce wastage. Begin by making a plan using a spider diagram to ensure you cover all the points. Write each point separately with an explanation. Include details of the effect on the environment, and the financial implications of food waste.

Now test yourself

TESTED

1 Give three reasons why food may be wasted. [3 marks]
2 Discuss why it is important for the environment to reduce food waste. [4 marks]
3 List four ways food wastage can be prevented in the home [4 marks]

8 Food provenance

Food origins

Growing food

Crops grown in the UK include:
- cereal foods such as wheat and barley
- vegetables such as potatoes and carrots
- fruit such as strawberries, raspberries, apples, plums and pears
- sugar beet
- oilseed rape for cooking oil production.

We cannot grow all the food we need in the UK, so some is imported from overseas.

How crops are grown

- Crop farming is **arable farming**.
- Weather and soil conditions determine which crops are grown.
- The process of growing crops is:
  1 preparing the soil
  2 sowing or planting seeds or seedlings
  3 watering
  4 controlling pests, fertilising crops and controlling weeds
  5 harvesting the crop.
- In England, southern and eastern areas are best for cereal crops as the land is flat and open.
- In Wales, farming tends to be sheep and dairy farming, as the land is more hilly and less suited to crop growing.
- Arable crops grown in Wales include wheat, barley, oilseed rape and maize for fodder. New potatoes from the Gower peninsula and Pembrokeshire are available early in the season.
- Farmers in Wales also produce more speciality products – Salt marsh lamb, Welsh Black Cattle beef and venison.

Figure 8.1 Crop production

Growing vegetables

● Many farms concentrate on one vegetable, for example potatoes or carrots, and provide different varieties for supermarkets and shops.

Growing soft fruits

● Soft, seasonal fruits, such as strawberries, are usually grown under polytunnels to provide protection from the weather and to control pests and reduce the need for pesticides.

Growing hard fruits

● Hard and stoned fruits, such as apples and plums, are produced on large- and small-scale farms in the UK.
● Kent, Worcester and Hereford are good fruit growing areas due to good soil and weather conditions.

Figure 8.2 A polytunnel

Types of farming

REVISED

● **Intensive farming:** uses pesticides and fertilisers to grow high-yield crops.
● **Organic farming:** food is produced as naturally as possible and has to stick to strict standards:
　○ No artificial chemical fertilisers or pesticides are used.
　○ Organic matter is used as fertiliser and wildlife is encouraged to control pests.
　○ Crop rotation, where different crops are grown each year in the fields, and animals are allowed to graze in those fields between crops, is encouraged.
　○ The fields are often left with no crops for a year (as fallow fields) to allow the soil to recover naturally.

How foods are reared

REVISED

● Animals raised for food include:
　○ pigs, cattle and sheep
　○ poultry, including chickens, turkeys, geese and ducks.

Factory farms

● **Factory farms** maximise the number of animals that are reared.
● Animals are raised in basic or poor conditions.
● They may have little space to move in.
● They are fed hormones to make them grow faster and drugs to reduce disease.
● Milk-producing cows can live in large sheds with no access to the outdoors.
● Chickens and turkeys are raised in huge indoor sheds, with no access to the outdoors.
● Egg-laying hens may be in caged conditions.
● This food will be cheaper.

Figure 8.3 Chickens reared in a factory farm

Organic farms

- The welfare of the animal is put first.
- Animals must:
  - have access to fields and be free-range
  - have living conditions that meet high welfare standards
  - be given a diet that is as natural as possible, with no added hormones to accelerate growth
  - be given drugs only to treat an illness.
- Organic food will be more expensive, as the animals take longer to develop naturally and are raised in more open areas, so not so many are produced.
- Currently, 35,000 hectares of land out of over 2 million hectares are farmed organically in Wales. Wales's first organic farm at Dol-y-Bont near Aberystwyth was started in 1948 by Dinah Williams and is now best known for the Rachel's Dairy business.

How foods are caught

REVISED

- Fishing methods differ in the UK depending on:
  - the type of fish being caught
  - the area that is being fished
  - whether it is small- or large-scale fishing
  - what technology the fisherman has.
- Types of fishing include:
  - **Trawling:** using a net that is dragged behind a trawler; different methods include pelagic, otter, beam and pair trawling.
  - **Purse seining:** drawing a huge net around a school of fish, for example tuna and herring.
  - **Dredging:** towing metal cages across the sea bed.
  - **Line fishing**: the most environmentally friendly form, as other fish are not harmed or killed.

Figure 8.4 Otter trawling, one of the methods of trawling

Farmed fish

- Fish are raised in fish farms, enclosed in huge cages in rivers, lakes or at sea. These fish include salmon, trout, cod and sea bass.
- These fish are fed wild fish, and often drugs and hormones to promote growth and reduce diseases.
- They can be raised in overcrowded conditions to keep the cost of production down.
- There are organic fish farms where the same standards as organic land farms will exist.

Lobster and crab pots

- Pots made of plastic, wood, rope and metal are baited with dead fish and placed on the sea bed.
- Lobsters and crabs enter the pot and cannot escape.
- The pots are then hauled onto the boat and the crabs and lobsters are removed.

Exam tip

You may be asked to give reasons why someone would choose to buy organic meat over factory-farmed meat. This would be in the form of an essay or discussion question. Remember to give a full explanation of the differences between the way the animals are reared, and to give detailed reasons why someone would choose organic meat, referencing the improved welfare conditions, the lack of added hormones and drugs and the improved taste of the final product. You can also include details of the cost of the meat, stating that even though the organic meat is more expensive, it is an ethical choice for the person.

Now test yourself

TESTED

1 Name four different fruit crops that are grown in the UK. [4 marks]
2 State two conditions that organic farms have to meet. [2 marks]
3 Explain how farm fish are reared, giving examples of fish reared in this way. [4 marks]
4 Give three reasons why someone would choose free-range, organic eggs over caged hen eggs. [3 marks]
5 Explain why it is important that smaller producers are making Welsh products that can be sold around the UK and abroad. [4 marks]

Food miles

- **Food miles** are distance that the food has travelled from the field to the plate it is served on.
- It also includes the distance the customer has travelled to buy the food.
- Consumers expect foods to be available all year round, so foods are transported from around the UK and from overseas.
- Food needs to be transported from where it is grown to the processing plant, and then to the shops and supermarkets where it is to be sold.
- Food is transported by sea, air and road.

Carbon footprint

REVISED

- **Carbon footprint** is the total amount of carbon emissions produced during the growing or rearing, processing and transportation of a product.
- Transporting food by air and road uses fuel, which produces polluting gases, such as carbon dioxide, from the fossil fuels being burnt.
- These gases contribute to **global warming**, which is the gradual heating up of the earth's atmosphere, seas and land.
- Reducing transport emissions can help to fight global warming; buying locally will help reduce transportation.

Figure 8.5 Transporting food

Buying food locally

Buying your food from local shops, farmers and producers:

- Supports the local economy and keeps the money in the area to spend on other local services
- Creates local jobs
- Is often less expensive
- Means food will be fresher and should contain more nutrients, as there is less time to lose valuable vitamins between harvesting and selling the produce
- Means it can be bought when it is in season and of better quality than food that has been transported thousands of miles.
- Is often safer, as there has been a minimal food chain, so there is less chance of contamination.

Exam tip

When you are asked about the benefits of buying local produce you need to discuss both local and global issues that will benefit from this. Make a list of all the benefits as a plan, and then expand on each point, so the answer is structured as an essay rather than a simple list. This will help you gain maximum marks.

Now test yourself

1 List three reasons why food is transported around the UK. [3 marks]
2 Explain what is meant by the term 'carbon footprint'. [4 marks]
3 Why is it important to buy local produce? [6 marks]

Packaging

- Most food we buy in shops is packaged in plastic.
- Loose foods, such as bread or fruit and vegetables, are usually placed in a paper or plastic bag when purchased.
- Food is packaged because:
  - it keeps the food safe and hygienic
  - it protects the food from damage and contamination, and stops people tampering with the food
  - it can extend the shelf life of the product
  - it can advertise the product
  - it provides information including ingredients, cooking and storage instructions and sell-by and use-by dates
  - it makes it easy to transport and store the food.

Chapter 10 includes details of the legal requirements of food labelling.

Exam tip

You may be asked to list advantages and disadvantages of certain types of packaging. Make sure you can provide some information about each point rather than just writing a list. Look at the marks available – if the question asks for you to **list** two advantages and two disadvantages, a simple answer will gain you the marks. If the question asks you to **discuss** the advantages and disadvantages, you will need to provide more information.

Why packaging is important for the manufacturer

- The manufacturer needs to provide information on packaging to comply with legal requirements.
- Colourful packaging can be used to attract the consumer.
- Packaging makes it easy for the manufacturer to store and transport the foods.

Types of packaging

● Glass – used for jams, jellies, sauces, pickled foods, instant coffee.

Table 8.1 Types of packaging

Type of packaging	Uses	Advantages	Disadvantages
Glass	Jams, jellies, sauces, pickled foods, instant coffee	● Transparent so the food is visible ● Recyclable ● Relatively inexpensive to produce ● Can be sterilised easily ● Waterproof ● Can be sealed so food has a long shelf life	● Very heavy to transport ● Breakable
Card or paper	Breakfast cereals, sugars, flour, teabags	● Cheap to produce ● Recyclable ● Light to transport ● Can be made into many different sized boxes easily	● Can contribute to global warming as trees are chopped down to produce it ● Not waterproof unless treated with a waterproof surface ● Easily damaged
Plastic	Yoghurts, milk, fruits and vegetables	● Light and flexible ● Cheap to produce ● Can be made into many different shapes and sizes ● Can be transparent or opaque ● Can be heat-sealed to extend shelf life ● Many plastics can be recycled ● Light to transport ● Waterproof ● Does not corrode	● Made using oil that contributes to global warming ● Some plastics are not recyclable so take hundreds of years to break down and cause pollution by filling landfill sites
Metal, foil or can	Canned vegetables and fruits, fizzy drinks	● Cheap to produce ● Sealed cans can extend shelf life to years ● Recyclable ● Foil containers can be placed directly into the oven ● Waterproof	● Uses valuable mineral sources ● Can be heavy to transport
Ovenable paperboard	Chilled and frozen ready meals	● Cheap to produce ● Can be heated directly in the oven ● Food labelling can be directly on the packaging ● Can sometimes be recycled ● Light to transport ● Can be moulded into different sizes and shapes	● Can be damaged fairly easily ● Not always recyclable as it is made from layers of different materials

Sustainability and food waste

Sustainability of food   REVISED

Sustainability means producing food in a way that it is not harmful to the environment, does not deplete natural resources and in a way that will continue to provide food for future generations. To do this we need to:
- avoid food waste
- reduce the impact on the local environment
- think about buying local and seasonal foods
- take into account farming and growing methods used to produce foods, including the use of fertilisers and chemicals.

The impact of food waste on the environment

- Billions of pounds worth of food is thrown away every year in the UK.
- Much of this ends up in **landfill** sites, where it rots and produces methane, which is a powerful greenhouse gas.

Local and global markets and communities   REVISED

Local markets

- **Farmers' markets** are now becoming more popular due to awareness of the benefits of buying locally.
- Buying from farmers' markets gives the advantages of buying locally, see above.

Global markets

- The range of foods available all year round in the supermarket is vastly different from 50 years ago.
- Food is transported from all over the world to supply demand from consumers.
- The world population is increasing, and millions of people do not have enough food to eat.
- The average income of the western world's population is high and means they can afford to spend more money on food, so competition for food is increasing.
- Food supply chains have increased in length. It can be common practice for packaged food to pass through several different countries before it ends up on the shelf of the supermarket.
- Consumers are being encouraged to buy locally to improve the sustainability of food.

Community farming

- **Community farming** is when communities join together to grow and produce organic, fresh food for their community.
- Land is often rented from a local farmer.
- Communities invest and work on the farm together, use the produce, and often sell the excess produce. The money raised from selling is reinvested in the project.
- Farming this way helps reconnect people with the land, encourages social gatherings and friendships and improves the health of the local community.

Food poverty

- **Food poverty** is when an individual or family cannot obtain enough healthy nutritious food.
- Families experiencing food poverty often rely on cheap foods that are high in fat and sugar do not provide the necessary nutrition for a healthy diet.
- People may sometimes have to miss meals, as they cannot afford to eat.
- There are many millions of people living in the UK who are in food poverty.
- Oxfam and Fare Share are two organisations that aim to reduce food waste and use excess food that is being disposed of by supermarkets to help feed people.
- Food poverty occurs for many reasons:
  - Low incomes mean people cannot afford to eat and pay household bills, such as fuel bills.
  - Food prices are rising faster than incomes.
  - Benefit changes may mean less money is available for food.
  - People are in debt that has to be repaid.
  - People are out of work and cannot afford food.
- Food banks have been set up in many towns and cities to provide basic food for people in food poverty.
- Food poverty is an increasing problem for low-income families in Wales. There has been an increase in the growth of food banks throughout Wales and in the number of families accessing crisis food provision.
- The main reasons people in Wales were referred to food banks were benefit delays, low incomes and benefit changes.

Exam tip

Questions on the sustainability of food could ask for a discussion essay on the importance of food sustainability at the current time. You need to refer to the concerns about global warming, food transportation, increased food poverty in the UK and how it is affecting the health of people. Include information about the increase in the global population, and the difficulties that may be experienced in the future about feeding the population of the world.

Now test yourself

TESTED

1 Explain how community farming projects can benefit local people. [4 marks]
2 Give three reasons why a working family could find themselves in food poverty. [3 marks]
3 Why is it important to encourage supermarkets to donate waste food to organisations like Fare Share? [4 marks]
4 Give three reasons why the number of people in Wales using food banks has increased so dramatically in the last three years. [3 marks]

Food security

Food security is defined by the World Health Organisation (WHO) as when 'all people at all times have access to sufficient, safe, nutritious food to maintain a healthy and active life', including 'both physical and economic access to food to meet their dietary needs as well as their preferences.'

Food availability

REVISED

- Enough food should be available all year round.
- The food source should be trustworthy.
- The country should be able to provide sufficient food both by importing and exporting food.

Food access

REVISED

- Food has to be available at an affordable price.
- There needs to be sufficient land available to grow foods.
- Suitable transportation has to be available to distribute the food.

Food use

REVISED

- Sufficient knowledge about a healthy diet, cooking skills and food-related illnesses should be available.

Key factors that influence food security

REVISED

- Disease.
- Safety of food sources.
- Reduced incomes.
- Increase in food prices.
- Increase in a country's population.
- Cost of food production for suppliers and the need to produce affordable food.
- Climate change and weather patterns, such as drought or monsoons.
- Amount of food grown or reared.
- Environmental issues such as food wastage.
- People's knowledge of healthy diets and food preparation.

> **Exam tip**
>
> A question could focus on food security becoming more of a global problem. It is important to include details of what factors affect food security and possibly give your own opinion on how the problem can be addressed in the future. Take each factor in turn and provide your own thoughts on how to solve the problem.

Now test yourself

TESTED

1 Define what 'food security' means. [2 marks]
2 List two points that food security is based on, and provide an explanation of each point. [4 marks]

9 Food manufacturing

Culinary traditions

British cuisine

- Traditional dishes are often hearty, filling meals that may include potatoes, for example: fish and chips, roast beef and Yorkshire pudding.
- Modern British cuisine has reinvented some of the traditional dishes to be cooked in a different way, such as adding new herbs and spices, presenting in a different way or adding unusual vegetables.
- Many famous chefs in Britain are influencing the way food is cooked and served.

Figure 9.1 A traditional roast dinner

Meal structures in Britain

- Traditionally, British people eat three meals a day: breakfast, lunch and dinner (the evening meal, which may be called tea or supper).
- A traditional British breakfast is a fried breakfast, but is now much more likely to be cereal or toast.
- Lunch is often a sandwich or wrap.
- The evening meal is traditionally 'meat and two veg', but now can be anything from a pasta dish to a take-away Indian or Chinese meal.

International cuisine

- Many different types of international cuisines are available in supermarkets and restaurants in the UK.
- The most popular ones are shown in the table on the next page.

Table 9.1 International cuisines

Cuisine	Example foods	Example dishes	Cooking methods and equipment
Indian	A wide range of herbs and spices, lentils, dhal, marinated meat, shallow-fried vegetables	Korma, jalfrezi, biryani, samosas, naan bread, chapattis	Karai, tava, tandoor
Chinese	Rice, noodles, bean sprouts, soy sauce, pork, eggs, ginger root	Chow mein, sweet and sour, spring rolls, Peking duck	Stir frying
Mexican	Spices, herbs, tortillas, beef, pork	Tacos, enchiladas, fajitas, chilli con carne	Stir frying, deep frying, baking
Italian	Olive oil, tomatoes, herbs (e.g. basil), cheeses (mozzarella, ricotta), salami, pepperoni	Pizza, pasta, risotto, panna cotta, biscotti	Pan frying, baking, simmering, grilling Wood-fired ovens for pizza
French	Meat, vegetables, herbs, cheese	Boeuf bourguignon, cheese soufflé, soupe à l'oignon, éclairs, tarte au citron	Braising, flambéing, sautéing, poaching

Meal structures in international cuisine

- Mediterranean countries, such as Italy and France have the same three-meal-a-day structure as Britain, but the main meal of the day is often lunch.
- Asian countries, such as China and India, will place a large number of dishes on the table at one time, giving a large choice of foods to diners. Breakfasts in India and China include rice and spicy dishes.

Figure 9.2 Meal structures vary in different countries

Welsh cuisine

REVISED

- Traditional Welsh dishes are historically based on the availability of local food. They include:
  - Cawl: a traditional lamb stew.
  - Bara Brith: a welsh tea bread.
  - Laverbread: made with seaweed that is washed and then cooked to a soft greenish-black paste.
  - Welsh rarebit.
  - Welsh cakes.

○ Glamorgan sausages: a vegetarian sausage made with leeks, cheese and potatoes.
- Other foods produced in Wales include meats such as Saltmarsh lamb, Carmarthen Ham and Welsh Black Beef, and cheeses such as Caerphilly and Tintern.

Now test yourself TESTED

1 Plan a three-course meal using traditional British cuisine. Identify where the main nutrient groups are found in your chosen dishes. [6 marks]
2 Explain why Asian cuisine uses a lot of rice. [2 marks]
3 Identify two popular Italian dishes that are eaten regularly by British people. [2 marks]
4 Explain why Welsh cuisine has different traditional dishes to the rest of the UK, and give examples of two of these dishes. [4 marks]

Primary stages of processing and production

- **Primary food:** a food that has been grown or reared and is not edible in its original state. It needs to be processed or prepared before it can be eaten. Examples include raw potatoes, wheat and milk.
- **Primary processing:** changing primary foods from their raw state into a product that can be eaten or used to make other food products. If can be simple, such as washing or peeling vegetables, or more complex such as milk pasteurisation.

Point of origin REVISED

- This is where the primary food has been grown or reared. For example, fruit is labelled with the country where it was grown.
- UK labelling laws require that the label for beef, veal, fish and shellfish, honey, olive oil, wine, most fruit and vegetables and poultry imported from outside the EU must show the country of origin.

Transportation of primary foods REVISED

- When foods are ready to harvest, or be slaughtered, they will be transported to a processing plant.
- Fruits and vegetables are be picked manually or by machinery and placed in containers for washing and sorting.
- Crops are harvested by machinery such as combines or tractors and transported to grain silos and stored for cattle feed, or sent to mills for processing into flour, oatmeal and other cereal products. Rapeseed is harvested and processed into oil, cattle feed and biodiesel.

- Cattle are put into transport trailers to travel to the slaughter house or abattoir.
- Refrigerated lorries can be used to keep foods fresh and prevent them going off.

Cleaning and sorting of raw foods

- Primary processing also involves cleaning and sorting crops.
- Potatoes will be put through a 'riddling' machine to remove stones, other debris and identify any damaged potatoes.
- They will then be washed, sorted into sizes and bagged.

Exam Tip

You may need to give examples of foods that undergo primary processing, and explain the process involved. Remember this is just the process the raw food goes through to make it edible, or able to be used to make a secondary product. Think about digging up a vegetable from the ground, or picking raw vegetables and think what you have to do to make it edible. An example would be picking pea pods, and removing the peas from the pod.

Now test yourself

TESTED

1 Give two examples of primary foods. [2 marks]
2 Explain what primary processing is and give an example. [4 marks]

Secondary stages of processing and production

- **Secondary processing**: when a primary food is changed or converted into an ingredient that can then be used to make a different food product. This is then called a secondary food product.
- Secondary processing can involve many different stages to produce the final product.

Table 9.2 Examples of secondary food products

Primary product	Secondary product
Wheat	Flour
Milk	Cheese and yoghurt
Fruit	Jam and jellies
Sugar beet	Granulated or caster sugar

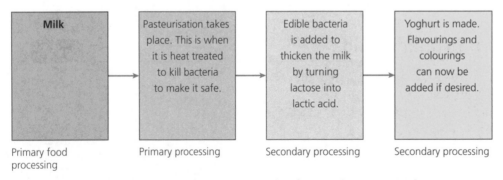

Figure 9.3 Turning milk into yogurt – an example of secondary processing

How food processing affects the sensory properties of an ingredient

REVISED

Many of these processes affect the appearance, smell and taste of a product.

Heating

All milk sold in the UK will have been heat treated to kill harmful bacteria.
- **Pasteurising** milk will not affect the taste or appearance.
- **Sterilising** milk (UHT milk) gives the milk a different taste because the heating causes a reaction between the sugars and the protein in the milk.

Cooling and freezing

- Cooling products does not affect taste, appearance or smell.
- Freezing products can affect the appearance. For example strawberries will freeze, but when defrosted will collapse – see Chapter 7.

Drying and curing

- Drying foods removes water, so the finished product is very different in appearance. An example is dried milk.
- **Freeze-drying:** food is frozen first and then placed in a strong vacuum. The water turn straight from ice to vapour. Foods that are freeze-dried include instant coffee, strawberries, apples and pears. Freeze-drying does not affect the taste so much as ordinary drying of foods.
- **Curing:** meat is treated with chemicals such as salt or sodium nitrate. Examples of these foods are bacon, salami and Parma ham. The flavour and colour of the meat is altered, as well as the feel and taste compared to the original product.

Figure 9.4 Cured meats

Smoking

- **Smoking:** the process of treating food by exposing it to smoke from burning or smouldering material, most often wood. Meats and fish are the most common smoked foods, though cheeses and vegetables can be smoked. This will alter the flavour and appearance of the food.

Fermenting

- **Fermenting:** foods produced or preserved by the action of micro-organisms (bacteria). Yoghurt and sauerkraut (pickled cabbage) are produced by fermentation. This process will alter the flavour, texture and appearance of the food, as well as the smell.

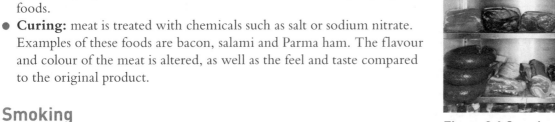

Now test yourself

TESTED

1 Explain what the term 'secondary processing' means. [2 marks]
2 List the stages involved in freeze-drying, and name two foods that can be freeze-dried. [4 marks]
3 Give two examples of methods of food processing where the overall taste and appearance of a food is not altered. [2 marks]

> **Exam tip**
>
> Make sure you can describe the process of secondary processing of one staple food, for example potatoes, rice or wheat, into a final product such as bread or cheese. Remember to include details of the primary processing stage, and then all the secondary processes.

Technological developments in food production

Using computers in manufacturing

- Computers can control the different processes accurately and improve the final outcome.
- The computer can also stop the whole assembly line at any given point. This will reduce wastage of food if there is a problem.
- Computers have also reduced the need for the number of manual workers in a factory, making cost savings for the manufacturers but reducing the number of jobs available for workers.

Table 9.3 Uses of computer in food production

Stage of processing	What the computer does	Advantages of using computers
Weighing of raw ingredients	Uses sensors to check accuracy of weight; controls the flow of ingredients	Saves time; ensures each product is identical in weight
Combining of ingredients to form the mixture	Controls the speed and the length of time of the mixing process	Saves manual effort; ensures all mixtures are correctly combined for an identical outcome each time
Dividing and portioning of the dough or mixture	Weighs each portion accurately or rolls to the same thickness	Ensures safety of workers during slicing and portioning; process is quicker; ensures each product is identical in size and shape
Baking of the food product	Controls the temperature of the ovens using accurate sensors; monitors the time so the product is not over- or under cooked	Ensures the safety of workers; each product looks identical
Packaging of the product	Checks for contamination by foreign bodies, such as metal detection	Ensures safety of the product

Fortifying and modifying foods to support better health

Fortification

- **Fortification:** adding nutrients to foods
- Nutrients lost during food processing may be replaced. This is particularly important if the food was a good source of a nutrient before processing, for example fortifying white and brown flour (see Chapter 1).
- Margarine has vitamins A and D added by law.
- Foods and drinks may have added nutrients to make them more appealing to the consumer, such as adding Omega 3 fatty acids, calcium or iron.
- In the UK cereals and flours are fortified with vitamins and iron and contribute to a healthy diet.
- Foods that are produced for vegans and vegetarians, such as soya products, are often fortified with Vitamin B_{12}.

Modified foods

REVISED

Modified foods include:

- foods that have added nutrients, such as yoghurts with added lactobacillus that claim to improve the gut flora in your body
- 'low fat' or 'low sugar' versions of existing products
- **genetically modified foods** that have had their DNA altered to increase or improve a favourable characteristic of a food product; this could be to improve the shelf life, resistance to disease or pests, or to increase the nutritional value of a food item.

> **Exam tip**
>
> If you have a question about the benefits of using a computer during the manufacturing of a product, think about each stage of the process, and write how much easier it can be for a programmed computer to control that stage.

Now test yourself

TESTED

1. Describe how a computer can be more beneficial when weighing out ingredients than a human operator on a production line. [2 marks]
2. Explain what fortification is, and give two examples of foods that have to be fortified by law in the UK. [4 marks]

Food additives

REVISED

- There are three different groups of additives:
  - **Natural:** obtained naturally from foods, for example beetroot juice for colouring
  - **Artificial:** made completely from chemicals
  - **Nature identical:** made chemically to be the same as the natural product (synthetic).
- All additives will be tested for safety and then allocated a specific number under EU law, so it can be identified in a product.

Table 9.4 Uses of additives

Type of additive	Why it is used	Examples of foods that contain this additive
Colourings (E100 numbers)	Enhances the attractive colour. Adds additional colour. Replaces colour lost during processing	Fizzy drinks, tinned peas, fruit yoghurts, sweets
Preservatives (E200 numbers)	Extends the shelf life of a product to stop it going off as quickly	Dried fruits, fruit preserves, cheeses, ham, sausages
Anti-oxidants (E300 numbers)	Stops food going brown by oxidation. Stops fats going rancid	Biscuits, jams, cut fruit, margarine, crisps
Emulsifiers and stabilisers and gelling agents (E400 numbers)	Helps foods mix together and prevents foods separating out. Extends the shelf life of baked goods. Gives foods a creamy texture	Low-fat spreads, mayonnaise, some baked goods
Acidity regulators and anti-caking agents (E500 numbers, but also include some E200 and E300 numbers)	Controls the acidity or alkalinity to a level important for processing, taste and food safety (not controlling pH levels can result in the growth of harmful bacteria)	Butter, pie fillings, fizzy drinks
Flavour intensifiers (E600 numbers)	Improves the flavour of a product or replaces flavours lost during processing	Yoghurts, soups, sauces, sausages

The advantages of using additives

- Food is kept safer for longer.
- Consistency of a product is maintained.
- Colours and flavours lost during processing are replaced.
- A product is made to be creamy or smooth, or more colourful, which improves it for a consumer.
- It may reduce the calorific value, for example when sweeteners are used to replace sugar.
- It can produce a range of product flavours, for example potato crisps.
- It gives a wide choice of foods for consumers.
- It can enhance the nutritional value, if extra nutrients are added.

The disadvantages of using food additives

- Some people can be allergic to certain additives, which may cause skin rashes or breathing difficulties.
- Some artificial colours may cause hyperactivity in children.
- Additives can be used to disguise low-quality ingredients or foods.
- Some additives have shown possible links to cancer if given in very large doses.
- As additives are used in so many foods that we consume, there is concern that eating so many will have long-term effects on health, but much research work is still being carried out.

Exam tip

Questions on additives may ask you to discuss the advantages and disadvantages of using food additives. Include information on how food additives are fully researched for their safety before being allowed to be used in food, and list the pros and cons, fully discussing each point for the top marks.

Now test yourself

1 Give two different types of food additives, and state their function in the food. [4 marks]
2 Explain the difference between natural and nature identical groups of additives. [4 marks]
3 List two advantages and two disadvantages of using additives. [4 marks]

10 Factors affecting food choice

Sensory perception

Sensory perception is the way we recognise flavour in food. Flavour is the sum of all the sensory stimulators, which are shown in Figure 10.1.

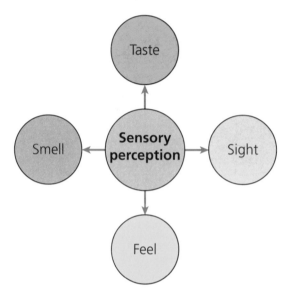

Figure 10.1 Sense involved in sensory perception

Taste

● Taste preferences develop through experience.

How taste receptors work

Taste buds are found on the upper surface of the tongue, on the soft palate, on the inside of the cheek, in the upper oesophagus, and on your epiglottis.

Your tongue's surface has tiny holes that let in the food after your saliva has dissolved it. The **taste receptors** are found at the top of the taste buds in these holes. Messages are sent to the brain to identify flavours.

There are five elements of taste:
1 **Sweet:** all sugary foods.
2 **Sour:** this is the taste that detects acidity. Found in foods such as lemons, grapes, oranges and sometimes melon.
3 **Saltiness:** any foods containing sodium, sodium chloride or potassium will give a salty taste.
4 **Bitterness:** common bitter foods include coffee, olives, citrus peel, dandelion greens and chicory.
5 **Umami:** this is a taste that is described as savoury or meaty. It can be tasted in cheese and soy sauce. It is present in fermented or aged foods, and in tomatoes, grains and beans. Monosodium glutamate, a food additive, produces a strong umami taste.

You taste these all over your tongue.

Smell

REVISED

- Eighty per cent of what we perceive of the flavour of food is due to smell, and only 20 per cent to just taste.
- The **olfactory system** is found in your nose. There are tiny hair cells in the olfactory system, which respond to particular chemicals and send messages to the brain to recognise smell. The nasal cavity and the mouth are connected – that is why both senses are involved in detecting flavour. Smell will increase saliva production, and release gastric juices and insulin into your digestive system.

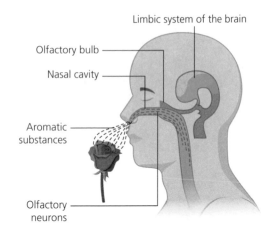

Figure 10.2 The olfactory system

Sight

REVISED

- Attractive food is more palatable and inviting to us, and we will often reject food that does not look attractive.
- Certain colours are associated with certain flavours in our brains, for example red colours are associated with fruit flavours.

Feel

REVISED

The part of the brain that deals with aroma or smell (the **olfactory cortex**) is linked to the part of the brain that deals with touch (the **somatosensory cortex**). You judge freshness of food on feel as well as taste, smell and sight, for example crisps are crunchy when fresh.

Exam tip

A question on this subject could ask for a list of how the four senses involved in taste work together. You would need to list the four senses, explain how they work and then describe how smell and taste work together, as the nasal cavity and mouth are connected in the body, and then explain how the part of the brain that deals with smell is linked to the part of the brain that deals with touch. Look at the marks offered and decide whether you need to list the five elements of taste as well.

Now test yourself

TESTED

1 Explain why someone with a head injury that has affected the part of the brain that deals with smell will not be able to taste food properly. [3 marks]
2 List all the elements of taste and give an example of a food from each one. [5 marks]

Tasting panels and preference testing

Sensory qualities of food

REVISED

Sensory qualities: the look or appearance, the smell or aroma, the taste and the texture and the sound of the food.

Sensory analysis: evaluating the sensory qualities of food by taste testing. By carrying out taste testing you are able to:
- identify key features of a product
- establish if there are any improvements needed
- determine if a product is suitable/acceptable
- compare similarities or differences in a food product
- test the quality of a food product.

Taste tests

REVISED

Ratings tests

In a **rating test**, taste testers could:
- Rate a food product for a particular characteristic such as how crispy it is or how salty it is, as well as give their opinion of the food product by rating it using a scale. This identifies if they really liked it or disliked it. It is called **preference rating**. Taste testers are asked to tick in a box to indicate how much they liked the food product.
OR
- Identify an order of preference for food products if they taste more than one. Taste testers are asked to rate them from the one they liked the best to the one they felt was the worst.

Profiling tests

A **profiling test** can be used to find out what people especially like about a food product. This type of test would help develop a profile according to a range of sensory qualities for example, chewiness, creaminess or saltiness. Taste testers would be asked to score the food product out of five.

Star profile

Tasters can use a **star profile** or **star chart** when they are assessing key sensory attributes of a product such as the appearance, texture, taste and aroma. The tasters rate the product according to the attributes. This can then be used to identify how the product could be improved.

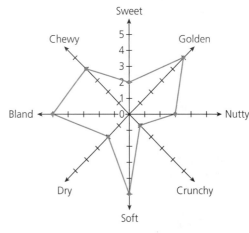

Figure 10.3 A star profile for almond cookies

Setting up tasting panels

REVISED

Industry

This is highly controlled to achieve reliable results. Aspects that are controlled include:
- the lighting and temperature of the area
- the use of individual booths to prevent any communication or influencing from other testers

- the use of trained testers
- the coding of food samples with random numbers
- the serving of food samples on identical size, colour and shaped plates or containers
- the temperature of the food being served to ensure it is identical for each tester
- the number of samples being presented to the tester
- the cleansing of the palette between the tasting of samples by the testers by drinking water or eating a plain biscuit/cracker
- the use of clear instructions to give to the tester
- the use of results sheets that are clear and easy for the testers to use.

School classrooms

- Ask as many people as you can so that you get a range of people's views on the food product.
- Ask people who fit into the target group, for example teenagers or vegetarians.
- Before the tasting session decide on the type of test that would be best for the tasters and draw up a tasting chart or a profile.
- Give clear instructions and try to seat the tasters so that they are not talking or sharing ideas with other testers to make sure that you have fair results.
- After the tasting session to analyse your results and identify where improvements or changes could be made.

Exam tip

If you are asked a question on setting up a taste test, make a short list of all the things that could influence a taste tester's decision. Try to plan a taste test that makes it as fair as possible, so that each point you have identified as being a possible bias will be eliminated. Think about the size of portion, random numbering, not letting people talk to each other and what sort of test you would use.

Now test yourself

TESTED ☐

1 Identify two forms of taste tests, and explain how they work. [4 marks]
2 List four things that you need to take into account to ensure a taste test is set up to be as fair as possible and give you the results needed to allow you to select your final dish. [4 marks]

Factors that affect food choice

Enjoyment

You will enjoy food more if:
- you are hungry: your taste buds become increasingly sensitive if you are hungry
- the food is palatable: sweet and high-fat foods have a high sensory appeal
- you are happy: your mood can influence enjoyment
- you are not feeling guilty about what you are eating.

Preferences

- Humans have a preference for sweet foods.
- Taste preferences and dislikes for certain foods also develop during experiences and through our attitudes, beliefs and expectations.

Seasonality

- Fresh fruit and vegetables are only available as freshly grown at certain times of the year when the crop ripens.
- Most foods are available all year round, as they are imported from other countries.
- **Seasonality** can also mean the times of year. We eat different foods in summer than in winter.

Figure 10.4 Vegetables are only available as freshly grown at certain times of year

Costs

- A person's income or the amount of money they have to spend on food will regulate their food choices.
- Low income groups are more likely to eat an unbalanced diet that is based on cheaper foods, which tend to be higher in fat and sugar.

Availability

- Seasonality is a factor, as discussed above.
- Access to transport and shops will restrict availability.
- If you live near a large choice of supermarkets, own a car and can drive, you will be able to buy everything you want, easily and quickly.
- If you live on a large housing estate, or outside a town and rely on public transport, it will be more difficult for you to buy and transport what you need. You may have to rely on small, local shops that do not stock such a large variety of products as large supermarkets.
- If you live in a village in the countryside, or on a remote Scottish Island, your choice will be much more restricted.

Time of day

- If you shop when you are hungry, you are more likely to spend more money.
- Many people only sit down and eat together in the evening, so choice of food at other meals may be a snack.

What activities you have planned

- If you are taking part in a sporting activity, your choice of food will reflect the activity.
- Many families have busy lives and do not eat together every day. This has an impact on choice of foods – whether ready meals are purchased, or fresh dishes are made and frozen or portioned in the fridge to be reheated.

Celebrations, occasions and culture

- Birthdays, weddings and anniversaries are often celebrated with a party, which involves buying and preparing special foods and dishes.
- Christmas is a huge family occasion when traditional foods like roast turkey and Christmas pudding are eaten.
- Many religions have their own celebrations and traditional foods that are eaten at certain times of the year.

Figure 10.5 A traditional Christmas dinner

Figure 10.6 Factors affecting choice

> **Exam tip**
>
> A question on this topic could be a longer essay-type question that asks for a discussion of factors that can influence and affect food choice. This will mean you have to make a plan to ensure you cover as many points as possible to access the top marks. Your plan could look like Figure 10.6.

Now test yourself

1 List four things that will affect your choice of food. [4 marks]
2 Explain why people on a low income are less likely to eat healthy foods, and suggest three ways they could improve their diet. [6 marks]

The choices that people make about foods, based on religion, culture or ethical belief, medical reasons or personal choices

Religions and cultures

Hindus, Muslims and Jews have dietary laws that dictate what can and cannot be eaten – see Chapter 4 for more details.

Ethical beliefs

Many people become vegetarian or vegan because of ethical beliefs. These are outlined in Chapter 4.

Medical reasons

- **Pregnancy:** there are certain foods that should be eaten to ensure the correct development of the baby, and certain foods that should be avoided.
- **Coeliac disease:** an allergy to gluten that requires careful selection of foods.
- **Type 2 diabetes:** a condition that must be carefully monitored and controlled by diet and insulin injections.
- **Cardiovascular disease:** a condition requiring a lower-fat diet.
- **Coronary heart disease:** a condition requiring a low-cholesterol diet.
- **Stroke:** requiring a similar diet to someone with coronary heart disease.
- **Obesity:** requiring lower-calorie foods to help with weight loss.
- **Allergies** and **intolerances:** conditions such as a nut allergy or lactose intolerance require avoidance of certain foods, or special products.

Personal preferences

These can be based on upbringing, religion, peer pressure, income or any of the other points discussed above.

Figure 10.7 Certain foods should be eaten and some should be avoided during pregnancy

How to make informed choices about food and drink to achieve a varied and balanced diet

Choose the correct nutrients

The main nutrient groups are:
- **Macronutrients:** proteins, fats and carbohydrates.
- **Micronutrients:** vitamins, minerals and trace elements.

A balanced diet will contain a variety of foods that provide the necessary nutrients.

Calculate energy requirements

- Work out your **BMR (Basic Metabolic Rate).**
- Calculate **PAL (Physical Activity Levels).**
- Use these to calculate your **EAR (Estimated Average Requirements)**. See page 37

Use government recommendations

- Use the Eatwell Guide for recommendations for healthy eating. See page 38
- Follow the guidelines for healthy eating.
- Use information about the nutritional needs of different age groups. See page 39

Use other specific websites and sources of information

- NHS website (www.nhs.uk).
- Consult organisations that are particularly dedicated to certain dietary conditions, such as the British Diabetic Association (www.diabetes.org. uk) or British Heart Foundation (www.bhf.org.uk).
- Look for dedicated charities that will provide suggested eating plans and recipes for specific conditions.

Calculate correct portion sizes

- **Portion size** can be calculated by looking at recipes, which usually give the number of servings for the finished dish.
- Check the packaging of a ready meal, which will state the number of servings per product.
- Calculate the total number of calories per portion by adding up the total calorie values of foods used in recipes and dividing by the number of portions you are serving. See page 48

Exam tip

A question could be based on the number of calories in a serving of a dish. Remember to add up the total number of calories and divide by the number of portions. If a food label of a ready-made product is shown, check for the number of servings and work out the number of calories per serving. See page 48

Now test yourself

1 Using the labels on the next page, calculate the total number of calories in a cheese sandwich using two slices of bread, 50 g cheddar cheese and 2 teaspoons of onion chutney. [3 marks]

Nutrition Information
(Typical values per 100 g)

Energy	416 kcal
Fat	34.9 g
(of which saturates)	21.7 g
Carbohydrates	0.1 g
(of which sugars)	0.1 g
Protein	25.4 g
Salt	1.8 g
Calcium	739 mg

Cheddar cheese

Nutrition

Typical values	100 g contains	Each slice (typically 44 g) contains	% RI*	RI* for an average adult
Energy	985 kJ	435 kJ		8,400 kJ
	235 kcal	105 kcal	5%	2,000 kcal
Fat	1.5 g	0.7 g	1%	70 g
(of which saturates)	0.3 g	0.1 g	1%	20 g
Carbohydrate	45.5 g	20.0 g		
(of which sugars)	3.8 g	1.7 g	2%	90 g
Fibre	2.8 g	1.2 g		
Protein	7.7 g	3.4 g		
Salt	1.0 g	0.4 g	7%	6 g

White bread

This pack contains 16 servings
*Reference intake of an average adult (8,400 kJ/2,000 kcal)

Values per 100 g		Values per teaspoon (10 g)
Energy	255 kcal	25 kcal
Fat	1.5 g	0.15 g
(of which saturates)	1.0 g	0.1 g
Carbohydrate	60 g	6 g
(of which sugars)	60 g	6 g
Fibre	1.5 g	0.15 g
Protein	1.0 g	0.1 g
Salt	1.0 g	0.1 g

Onion chutney

Figure 10.8 Nutritional information for the elements of a cheese sandwich

Food poverty in Wales

- Almost one in four people in Wales lives in poverty, which means they get less than 60% of the average wage. That is about 700,000 people.
- For more on the food and nutrition strategy for Wales, see Chapter 4: Plan Balanced Diets.

Food labelling

From December 2014 the following information has to be on a food label by law:

- the **name of the food** – for example 'wholemeal bread' or 'cottage pie'
- a **best-before** or **use-by date** (see Chapter 7)
- **quantity information** – must be in grams, kilograms, millilitres or litres on the labels of packaged food over 5 g or 5 ml
- a **list of ingredients** (if there are more than two) – must be listed in order of weight, with the heaviest listed first
- **allergens** – if the product contains any allergens it must say so clearly on the label and be listed in the ingredients; there is a list of allergens that you must mention if used
- **name and address of the manufacturer**, packer or seller – contact details needed in case there is a problem with the food
- **lot number** of the food – means it can be traced back to the production date
- any special **storage conditions** – e.g. 'store in the refrigerator'
- **instructions for cooking** if necessary
- **country of origin** – if the following products in the food have been imported from outside the EU: veal, beef, fish, shellfish, honey, olive oil, wine, most fruit and vegetables, poultry
- a warning if the product contains **GM ingredients** unless they are less than 0.9% of the final product
- a warning if the product has been irradiated
- the words '**packaged in a protective atmosphere**' if the food is packaged using a packaging gas – this could be in Modified Atmospheric Packaging, for example
- any necessary **warnings** – there is a list of foods from government legislation showing which ingredients and chemicals must be listed, and the actual wording that has to be added to each ingredient.

From December 2016, the law states that mandatory nutritional information must be included on food labels.

Other information that may be on the labels

The following information could be on a label:

- nutritional information:
  - energy per 100 g and per portion
  - **Reference Nutritional Intakes (RNIs)**
  - the percentage of the RNI that is used for fats, saturated fats, sugars and salt
  - **traffic light symbols:** used to indicate whether a product is high (red), medium (amber) or low (green) in fats, saturated fats, sugars and salt. These also show the amount of nutrients in a portion of food and drink and the percentage of your RNI that a portion of the product will provide.

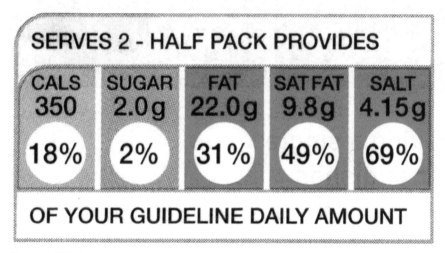

Figure 10.9 Traffic light symbols

- 'may contain nuts' or other ingredients that may be allergens
- nutritional and health claims: have to have been verified by EU law – examples:
  ○ Sugar free (must contain less than 0.5 g of sugar per 100 g of product)
  ○ Low fat (must contain less than 3 g of fat per 100 g of product)
  ○ High in fibre (must contain more than 6 g of fibre per 100 g of product)
- **E-numbers**, **anti-oxidants** and **preservatives**
- flavourings, flavour enhancers, sweeteners, emulsifiers and gelling agents
- marketing terms such as 'pure, natural'
- vegetarian and vegan labelling
- 'made with real fruit' or 'contains real fruit juice'
- 'wholegrain' or 'made with wholegrains'
  ○ 'wheat flour' or '100% wheat' – you need to look for wholewheat flour
  ○ 'multigrain' – means there are several types of grains, not necessarily wholegrains
  ○ 'wholegrain' – check the ingredients and avoid words like 'bleached' or 'enriched'; 100% wholegrain is the best.
- celebrity endorsement: very popular to persuade people to buy
- cartoons or free gifts: to appeal to children
- 'new recipe' or 'improved recipe'.

> **Exam tip**
>
> A question could ask for reasons why you think a manufacturer would put extra information on a food label that is not required by law. You would need to give several examples of what this information could be and give reasons as to why you think these are included. The main reason is to persuade people to buy the product, but also it gives customers information to allow them to make a reasoned choice and provides nutritional information.

Now test yourself

TESTED

1 List four pieces of information that have to be on a food label by law. [4 marks]
2 Why do you think some manufacturers use cartoon characters to advertise food for children? Explain why you think it is a good or a bad idea to do this. [5 marks]
3 There are suggestions that information showing how many teaspoons of sugar are contained in each product should be added to a label. Do you think this would help prevent childhood obesity? Give reasons as to why you think this would either be helpful or not helpful. [4 marks]

11 Preparation and cooking techniques

Planning for cooking a single dish or a number of dishes

- Look at the task and select an appropriate dish or dishes. For example, you may be planning and cooking for a vegan.
- Look at the amount of time you have to prepare and cook the dish or dishes. Make a **time plan** to ensure you can complete the work and clear up in the time given.
- Show **health and safety points** on your time plan.
- If you are cooking several dishes you need to make a **dovetailed** time plan, planning the dishes in the correct order, so the longest one is started first.
- Check the **cost** of the ingredients and make sure the dish(es), is/are affordable and fit(s) in with your family's budget.

Making and cooking the dish or a number of dishes

REVISED

- Follow the time plan.
- Taste and season the dish or dishes during cooking.
- Note any extra instructions, such as glazing the dish.
- Pay attention to any instructions that will affect the outcome, such as correct thickness of pastry during the rolling out.
- Check the dish for the correct cooking time and taste or test food during the cooking process to ensure it is correctly cooked and seasoned.
- Think about the presentation of the dish. Use garnishes to make it as attractive as possible.

Evaluating the dish

REVISED

- Use **sensory descriptors**.
- Ask other people to comment on your dish. You should comment on other people's work.
- Suggest any improvements that will make the dish taste or look better.

Now test yourself

TESTED

1 Explain how creating a time plan for your cooking session can help to ensure a successful outcome. **[4 marks]**
2 List three things to consider when selecting dishes for a given task. **[3 marks]**

> **Exam tip**
>
> A question on this topic could ask for a number of factors that you need to take into account when selecting a dish for a particular task. Make sure you list each point and write an explanation of the point. For example, 'A vegan does not eat any animal products, so I must ensure that any sauces or added ingredients do not contain any animal products, such as gelatine or egg white.'

Now test yourself answers and quick quizzes at **www.hoddereducation.co.uk/myrevisionnotes**

Preparation of ingredients

Knife skills

REVISED

Using different knives:
- vegetable knife: used for small fruit and vegetable preparation, using the **bridge hold**
- cook's knife: used to carry out the **claw grip**, which is used for larger fruit and vegetables
- filleting knife: used for meat and fish
- chopping knife: used to chop, slice, shred and dice vegetables, or chop herbs – includes julienne, dicing and cutting batons.

Figure 11.1 The bridge hold

Figure 11.2 The claw grip

Preparing fruit and vegetables:

REVISED

This includes:
- mashing, shredding, scissor snipping, scooping, crushing, grating, peeling and segmenting
- de-skinning, de-seeding, blanching, shaping, piping, blending and juicing fruit and vegetables
- preparing garnishes.

All of these must be done while demonstrating the skills of controlling enzymic browning.

Prepare, combine and shape

REVISED

This includes:
- rolling, wrapping, skewering, mixing, coating and layering meat, fish and alternatives, for example making fajitas
- shaping and binding wet mixtures, such as fish cakes and meatballs.

All of these must be done while demonstrating the prevention of cross-contamination and showing that you can handle high-risk foods safely.

Tenderise and marinate

REVISED

This includes:
- showing how acids **denature** proteins by using a marinade, for example marinating meat prior to barbequing it
- showing how **marinades** add flavour and moisture while preparing meat, fish, vegetables and alternatives.

Select and adjust a cooking process

REVISED

This includes:

- selecting and adjusting the length of time to suit the ingredient
- understanding that the instructions given in a recipe for the length of cooking time will be an estimate, based on a particular size or portion of meat, fish or alternative protein source
- making sure you know how to adjust the cooking time to make sure the portion is not undercooked or burnt
- checking during the cooking time to see if a meat portion is cooked by using a meat thermometer, or by inserting a knife to see if any blood still leaks out
- understanding that fish takes much less time to cook, so this will need to be checked carefully by gently poking a fork into the flesh; it should be opaque and flake away from the fork.

Figure 11.3 Fish must be checked carefully

Weigh and measure

You will need to be able to weigh and measure dry and liquid ingredients accurately.

Preparation of ingredients and equipment

In order to prevent food sticking to tins it may be necessary to grease, oil, line or flour cake tins or baking sheets before you use them.

Use of equipment

Many recipes require the use of blenders, food processors, food mixers and microwave ovens as part of the preparation or cooking process.

> **Exam tip**
>
> Questions on preparation techniques are likely to be included with other topics. For example, you could be asked how to make a tough piece of meat tenderer, so you could explain the marinating technique and how it makes the meat tenderer.

> **Now test yourself**
>
> TESTED
>
> 1 Explain how you would safely dice an onion. Describe which knife you would use, and how you would hold the onion to prevent any accidents. [3 marks]
> 2 Why is it important to prepare a cake tin before adding the mixture for a rich fruitcake? Explain what you would do to the cake tin. [2 marks]

Cooking a selection of recipes

Water-based methods using the hob

REVISED

- **Steaming:** the cooking of prepared foods by steam (moist heat) under varying degrees of pressure. High-pressure steaming is carried out in a pressure cooker. In low-pressure steaming, food may be cooked as follows:
  - ○ Indirect contact – the food is placed in a perforated container above the boiling water and the steam rises through the perforations and cooks the food, for example steaming vegetables or fish.
  - ○ Direct contact – place a container into the boiling water, for example cooking a steak and kidney pudding.
- **Boiling and simmering:** foods are placed directly in boiling or simmering water in a saucepan.

Figure 11.4 A steamer

- **Blanching:** vegetables are blanched before freezing and to partly cook them so that they can be reheated and served later.
- **Poaching:** cooking in liquid with a temperature ranging from 60 °C to 82 °C, usually eggs or fish.

Dry heat and fat-based methods using the hob

- **Dry-frying:** placing an ingredient such as mince or bacon, which contains fat, into a dry frying pan. The fat in the meat melts and can then cook the meat.
- **Pan-** or **shallow-frying:** a small amount of fat or oil is put into a frying pan, the oil is heated and then the food is placed in the hot fat and cooked.
- **Stir-frying:** usually done in a wok. It uses very little oil. The wok is heated to a high temperature and the food is added and cooked quickly, as it is shredded into thin pieces.
- **Deep-frying:** Food is totally immersed in hot fat.

Grilling

- You can use the grill for meats, fish, vegetables and alternative sources of protein like halloumi cheese. You can also toast seeds and nuts under the grill.

Using the oven

- **Baking:** pastry-based products, cakes, biscuits, breads, fish and other protein-based foods.
- **Roasting:** meats, fish and vegetables.
- **Casseroles** and **tagines:** slow cooking in a liquid.
- **Braising:** usually frying a product to brown it, such as a cheaper cut of meat, and then putting it in a casserole dish to stew gently in the oven.

Making sauces

Heated sauces show **gelatinisation**, where the starch granules swell as they are heated. When the liquid boils they burst and absorb the liquid, thickening the sauce. The more starch you add to the liquid, the thicker the sauce will be.

You need to know the following sauces and how to make them:
- **blended** white sauces, for example all-in-one sauce and roux-based sauce
- **infused** sauces with a roux base such as velouté and béchamel
- **reduction** sauces, for example pasta sauces, curry sauce, gravy and meat sauce
- **emulsion** sauces, for example mayonnaise and Hollandaise.

Figure 11.5 A velouté sauce

Set a mixture by removal of heat (gelation)

- **Gelation:** to solidify by freezing or chilling.
- If a mixture is chilled or frozen, it will set. You can see this happening in a cheesecake recipe, or in blancmange or custard on a trifle. Ice cream solidifies as it is frozen.

Set a mixture by heating (coagulation)

REVISED

- An example of this is when eggs **coagulate**, by protein **denaturation**, when the mixture is heated, for example choux buns or a quiche filling.

Use of raising agents

REVISED

These could include:
- incorporating **air**, which will expand when heated and become trapped in a mixture when the gluten sets, such as in a creamed sponge mixture or a whisked meringue mixture
- using a **chemical raising agent**, such as self-raising flour, baking powder, bicarbonate of soda or yeast
- using **steam** in a mixture, such as when making choux buns or chocolate éclairs.

Make a dough

REVISED

Making pastry, bread or pasta will show the following:
- **Shortening:** the ability of fat to coat the flour particles, forming a waterproof layer and preventing the gluten forming long strands, so the final product has a crumbly, 'short' texture. This happens when you create a 'rubbed-in' mixture for shortcrust pastry or shortbread.
- **Gluten formation:** gluten in the flour is stretched during the kneading of bread dough and forms long strands of protein, which will act as the structure of the bread. This hardens as the bread is cooked and traps the air to form the final product.
- **Fermentation:** yeast in bread dough begins to grow, feeding on the sugar, and carbon dioxide is released causing the bread to rise.

Figure 11.6 **Preparing palmiers**

Shaping a dough

You will need to be able to:
- roll out pastry for a product or recipe
- roll and make fresh pasta
- make a shaped bread roll, loaf of bread or flat breads
- make pizza or calzone
- layer puff pastry and create a final outcome such as palmiers
- glaze a product with an egg wash, to give a golden brown finish; this could be a pastry product such as sausage rolls
- pipe a product such as choux pastry into éclairs.

Test for readiness

REVISED

- **Using a temperature probe:** check that any dish containing beef, lamb, chicken or pork has reached 72 °C for at least two minutes to ensure harmful bacteria has been killed.
- **Using a knife or skewer:** check if a cake is cooked by inserting a knife or skewer and seeing if it comes out clean, with no mixture stuck to it. If you stick a knife or skewer into the thickest part of a chicken leg and the juice that runs out is clear, with no blood, it means the chicken is cooked. Vegetables will be soft when a knife is stuck into them.

- Using a **finger** or **'poke' test**: This is often used when you are cooking a cake to see if the cake springs back when lightly touched. It is also used when kneading bread dough to see if it is elastic and springs back.
- Using a **'bite' test**: This is usually done with cooking pasta, to see if the pasta is cooked properly.
- Using a **visual colour check**: Often you can tell from the colour of a product if it is cooked. Biscuits, cakes and pastry products will turn golden brown when cooked.
- Using a **sound check** to see if a bread roll or loaf is baked, as it will sound hollow when the base is tapped.

Judge and manipulate sensory properties

REVISED

- Taste and season during the cooking process
- Change the taste and aroma through the use of infusions, herbs and spices, paste, jus and reductions
- Change the texture and flavour using browning (dextrinisation) and glazing; adding crust, crisp and crumbs

Presenting a selection of recipes

Garnishes

REVISED

A **garnish** is an item or substance used as a decoration or embellishment accompanying a food dish, to give added colour or flavour. Garnishes are usually edible.

- Fruit garnishes:
  - citrus fruit flower
  - strawberry flower
  - strawberry fan.
- Vegetable garnishes:
  - carrot flower
  - cucumber curl
  - potato rose
  - cherry tomato flower
  - tomato rose.

Figure 11.7 Citrus fruit flower

Figure 11.8 Strawberry fan

Figure 11.9 Carrot flower

Figure 11.10 Potato rose

Figure 11.11 Tomato rose

Piping

You can pipe mashed potato, cream and icing onto dishes and cakes.

Glazing

REVISED

- You can **glaze** dishes before they are cooked to give a shiny, golden brown finish, or you can glaze food after cooking, for example with melted butter to give a shine.
- You can use jelly or aspic to coat savoury foods to give a clear, shiny coating. Apricot jam is sometimes used to give a shiny coating to desserts and cakes.

Now test yourself

TESTED

1 List two dishes where you would use a knife to ascertain if the dish was cooked. Explain how you will know if the dish is cooked. [4 marks]
2 Why is it important to garnish a dish? Give four different ways you could improve the appearance of a completed dish. [6 marks]

Exam tip

It is not likely that a direct question on cooking techniques will be asked, but you need to know how to explain what happens during the cooking techniques listed above.

Working safely

Prevention of cross-contamination is vital during food preparation, as is keeping yourself safe.

Personal hygiene

REVISED

- **Tie long hair back:** this prevents your hair falling into food and contaminating it.
- **Wash your hands:** always wash your hands before you start cooking, after handling raw meat and fish, after using the toilet and after blowing your nose or sneezing on your hands. See Chapter 7 for more details on the risks of cross-contamination from not washing hands.
- **Remove any jewellery:** rings on your fingers can trap bacteria underneath. All rings should be removed before cooking.
- **Wear a protective, clean apron:** your clothes have bacteria on them. You need to protect the food from these bacteria.
- **Keep your nails short and clean,** with no nail varnish: nails can trap bacteria; flakes of nail varnish can fall into food, contaminating it.
- **Cover any cuts or boils** with blue kitchen plasters: cuts and boils can contain harmful bacteria. Blue plasters are used because if they fall into the food they can be easily spotted.
- **Do not cough or sneeze over food:** this will contaminate the food with bacteria.
- **Do not cook if you are unwell,** particularly if you have a stomach upset: the bacteria causing the illness could be transferred into the food and spread to other people who eat the food.

Store foods in the correct places

REVISED

- Frozen foods need to be put in the freezer and not allowed to defrost.
- Chilled foods need to be in the refrigerator, on the correct shelves.

- Dry goods need to be stored in sealed containers, off the floor of the larder or storage room.
- Tinned foods need to be stacked on shelves off the floor.
- Use foods in date order, check sell-by or use-by dates.

Use all equipment safely

REVISED

Knives

- Use the correct knife and grip when preparing food.
- Use different knives for raw and cooked food to avoid cross-contamination.
- Blunt knives are more dangerous than sharp knives, as they need more pressure and can slip and cut.
- Do not put knives into a washing-up bowl and leave them, as people can easily put their hands in the water and cut themselves.
- Keep knife handles clean and grease free so they do not slip when using them.
- Never walk around the kitchen holding a knife with the blade pointing outwards.
- After washing and drying the knife, return the knife to the correct storage place.

Figure 11.12 **Using the correct knife is important when preparing food**

Electrical equipment

- Check all machinery is in good working order, with no frayed cables or loose plugs.
- Do not handle electrical equipment with wet hands or you could get an electric shock.
- Never drape electric cables in or over water or wet patches on work surfaces.
- Only one person uses the equipment at a time.
- Keep your hands away from the moving parts.
- Wash blades and beaters, dry carefully and replace onto the equipment.
- Place the equipment on a work surface so it is stable and not liable to topple or fall.
- Turn all equipment off and unplug when you are not using it.

Hot ovens, hobs and equipment

- Use oven gloves or oven cloths, not tea towels, to place food in and remove food from the oven to prevent getting burnt.
- Turn pan handles away from the other hot plates if they are being used, so the handles do not get hot and burn you.
- Take care when moving hot pans or dishes; ask people to move out of the way if you are sharing a cooker or hob.
- Never put hot dishes directly onto a work surface or chopping board; use a cooling rack or trivet.
- Do not leave oven doors open or hot plates turned on or lit when you are not using them. Do not leave a grill unattended.
- Do not clean the oven or hob while it is still hot.

Spillages

- Mop up any spillages on the floor or work surfaces immediately to prevent people slipping or burning with hot liquids.
- Use the appropriate floor or dish cloths to prevent cross-contamination.

Cleaning equipment

- Clean up as you go while cooking so that there are not piles of dirty dishes over your work surface or in the sink area.
- Wipe surfaces down regularly to prevent contamination.
- Use the appropriate cloths and cleaning fluids for each area. Do not mix cleaning fluids.
- Ensure food wrappings and peelings are cleared into the correct bin before you start preparing food.
- Keep bins covered. If you have to touch the bin to remove the lid, make sure you wash your hands before preparing your food.

> **Exam tip**
>
> Questions on this topic are often in the form of asking for a number of points on a particular subject, for example a question could ask for five personal hygiene rules. Make sure you only discuss personal hygiene; do not put points in about food storage or food preparation.

Now test yourself

TESTED

1 List four safety points that need to be observed when working with electrical equipment. [4 marks]
2 Explain why someone who is suffering with a stomach upset should not be working in a kitchen preparing food. [4 marks]

Using sensory descriptors

Sensory descriptors are used to describe the flavour, aroma, texture and appearance of your finished dish.

Describing taste, flavour and smell

REVISED

- acid
- aftertaste
- astringent
- bitter
- bland
- burnt
- creamy
- dry
- hot
- fatty
- metallic
- old
- piquant
- pungent
- salty
- sharp
- sickly
- soggy
- sour
- spicy
- tangy
- tart
- zesty.

Describing appearance

REVISED

- appetising
- attractive
- bubbly
- cellular
- clear
- cloudy
- coarse
- colourful
- colourless
- crumbly
- dry
- fat
- flat
- foamy
- fresh
- greasy
- grained
- healthy.

Describing texture

REVISED

- adhesive
- bouncy
- brittle
- bubbly
- cellular
- chewy
- close consistency
- coarse
- cohesive
- cold
- crisp
- crumbly
- crunchy
- crystalline
- dry
- elastic
- effervescent
- fibrous
- firm
- fizzy
- flabby
- flaky
- foamy
- juicy
- lumpy
- moist
- mushy
- pulpy
- powdery
- slimy
- smooth
- soft
- soggy
- spongy
- springy
- sticky
- stiff
- stretchy
- stringy
- syrupy
- tacky
- tender
- thick
- thin.

12 Developing recipes and meals

The influence of lifestyle and consumer choice when adapting or developing meals and recipes

REVISED

- **Religious beliefs:** Certain religions have restrictions on what they can eat, or they can only use food that has been slaughtered in a certain way. You can read about these in Chapter 4.

Figure 12.1 A Halal butcher's shop

- **Ethical choice:** Some people choose not to eat animal foods. See Chapter 4 for details.
- **Illness:** Some illnesses require special diets.
  - ○ Type 2 diabetes: Eating a balanced diet is important, as well as eating regular meals to maintain a steady blood sugar level.
  - ○ Cardiovascular disease and coronary heart disease. A healthy, balanced diet is the best way to lower or maintain lower cholesterol conditions, reducing the risk of heart attack or stroke.
  - ○ Coeliac disease: Sufferers have to avoid any foods containing gluten, so using special products that are gluten-free as substitutes will make the food suitable.
- **Family lifestyle patterns:** Working parents may have limited time to spend cooking meals and rely on ready-made processed food. You can batch-cook at the weekend, and freeze portions of food that can be reheated on a week night and served with freshly cooked vegetables, or use quick recipes from a cookery book or website.
- Families that do not often eat together can still have healthy meals if they freeze individual portions of food in the freezer. This means each person can reheat the food when needed.
- **Limited budget:** There are many ways to save money, and still provide a healthy, balanced diet:
  - ○ Look for products that are on offer in a supermarket.

> **Exam tip**
>
> A question on this topic may ask you to alter a given recipe to make it suitable for a particular target group. Make sure you explain what the necessary adaptations are in detail, and why you are making this adaptation. For example, if you were changing an ingredient to make the recipe suitable for a vegan, you would state that 'a vegan eats no animal produce, so I am changing the cheddar cheese for a vegetarian cheese substitute that is made with no milk or animal rennet to set it, making it suitable for a vegan.'

○ Buy products from the 'reduced' section, where they are nearing their sell-by or use-by date.

○ Buy 'own brand' products rather than premium brands.

○ Shop at cheaper supermarkets, where good quality products are available at much lower prices.

○ Use canned or frozen fruit and vegetables instead of fresh.

○ Shop at markets rather than supermarkets as you can often buy smaller quantities of ingredients rather than a large pack.

○ Use leftovers to make another meal.

Now test yourself

TESTED

1 Suggest three ways you could reduce the cost of a recipe for a beef lasagne for a family on a low income. [3 marks]

2 Explain why you could not use meat from a normal butcher for someone who is a Muslim. [2 marks]

3 Discuss ways in which a family on a low budget could still eat a healthy, balanced diet. [6 marks]

Adaptations to recipes to address current dietary advice

REVISED

Current government advice includes using the **Eatwell Guide** to plan your meals. You can read more on the Eatwell Guide and the eight rules for healthy eating in Chapter 4.

Figure 12.2 **Base meals on starchy foods**

There are also **RDI (Recommended Daily Intake)** amounts for all the main macronutrients and micronutrients. You can read more about RDI amounts in Chapter 4.

Wales also has further suggested government advice, including:

● Change4life Wales: aims to help people make small, incremental changes to their lifestyle to improve their health and well-being.

● Food for Wales, Food from Wales 2010/2020: Food Strategy for Wales.

Exam tip

A question on current government advice could ask for an explanation of the Eatwell Guide, and why each segment is a different size. You would need to explain why the recommended segment sizes are for a third of the plate being the fruit and vegetable section, and a further third being starchy foods. Remember to discuss the nutrients found in fruit and vegetables, and the importance of eating starchy foods to provide slow-release energy throughout the day. Also refer to the smaller segments on the plate, with protein foods including beans and pulses for non-meat-eaters.

Now test yourself

TESTED

1 List four of the recommended eight government rules for healthy eating and explain the importance of each one. [4 marks]

2 Discuss why the blue and purple portions on the Eatwell Guide are the smallest. [4 marks]

3 Using the Eatwell Guide, plan a meal that would show the correct proportions of nutrients recommended. Explain where each of the nutrients is found in your meal. [6 marks]

Considering nutritional needs and food choices when selecting recipes

- When selecting recipes for people with **intolerances** or **allergies** check ingredients carefully, and substitute any ingredients that could be dangerous or cause illness or discomfort. Check with the person or people that you are cooking for before you begin planning the dish as to whether they suffer from any allergies or intolerances.
  - **Nut allergies:** can be life-threatening. A person can have a mild allergic reaction, but can also suffer anaphylactic shock which can kill them.
  - **Lactose intolerance** will cause discomfort for the sufferer, and many lacto-free alternatives are available.
  - Some people can be allergic to **eggs** or **shellfish**.
  - There is a government list of allergens that must be included on the labels of food products.
  - **Lifestyle choices:** Vegetarians and vegans will require adaptations to recipes.
- **Cooking processes and methods:**
  - Conserving nutritional content: check how certain nutrients are destroyed by some cooking methods. See Chapter 6 for more details.
  - You may also be cooking for elderly people who have difficulty chewing, so choice of cooking methods or softer foods is important.
- **Portion control** for younger children and someone trying to lose weight needs to be considered.
- Someone with **high energy needs**, such as a sports person or athlete will need specific types of meals to maximise their energy requirements.

Figure 12.3 A meal suitable for someone who has difficulty cutting food

---

Exam tip

A question on this section could ask for a suggested adaptation to a recipe to make it suitable for an elderly person who has difficulty cutting food because of arthritis, and difficulty chewing.

You would have to suggest a meal that contains softer food that is easy to cut and chew, for example a piece of poached fish with mashed potato and sliced, steamed cabbage. Think about the nutritional content of the meal, cooking methods to conserve nutrients and also the appearance of the meal, to give an attractive, flavoursome, colourful dish that will encourage the elderly person to eat.

---

1 Explain why it is important to check ingredients for a recipe if you are making a meal for someone with a nut allergy. [2 marks]
2 Give two ways you could cook broccoli to conserve the Vitamin C content. [2 marks]
3 Suggest three ways to increase the energy provision of a diet for a sports person. [3 marks]

Reviewing and making improvements to recipes

REVISED

- Evaluate your cooking using sensory descriptors. This will mean you will be able to suggest improvements for that dish or meal if you were going to cook it again.
- Use tasting tests to find out people's opinions. This can help you to suggest alterations and improvements.
- Look at a chosen recipe and identifying any changes or improvements that can be made during the planning process before you make it. Consider:
  ○ Are you using appropriate ingredients? If the dish or meal is for a particular age, dietary condition, illness or intolerance, have you checked that all the ingredients are suitable for that person? If not, what changes do you need to make?
  ○ Are you using the correct process and cooking method? Will the preparation and cooking processes preserve nutrients? Will the food discolour while it is waiting to cook? Is it going to be cooked in time? Can it be reheated safely? Will it look attractive?
  ○ Are the portion sizes correct? Who is the food being cooked for? Have you calculated the energy content if the food is for a low-calorie or high-energy diet?

Manage the time and cost of recipes

REVISED

- Make a **time plan** prior to cooking: to see if the dish or dishes you have chosen to cook fit in the amount of time you have. The dishes will need to be **dovetailed** to show that you are using your time efficiently and effectively.
- Consider the **cost of ingredients**: The cost of recipes is also important, not only if the brief requires a low-cost meal, but also to ensure your own family is not spending excessive amounts of money. It is possible to select less expensive alternatives by choosing less expensive varieties of fish and meat, 'own brand' packaged goods or 'wonky' fruit and vegetables.

Using testing and sensory evaluation skills

REVISED

- Check when food is cooked: Use a temperature probe, a knife or skewer, a finger or the 'poke' test, a 'bite' test, or a visual colour check or sound check.
- Taste your food for the correct seasoning during cooking and adjust to produce the most flavoursome outcome.
- Adjust cooking times. For example, if you change the portion size of the recipe, by reducing the servings from four to two, as well as halving the quantities in the recipe, the cooking time may need to be reduced.

- Add more colour or texture to the dish. This can be achieved by adding colourful side dishes, accompaniments or a simple garnish.
- Present your food attractively, using garnishes in the most aesthetically pleasing way.

Figure 12.4 Checking food has been cooked

Now test yourself   TESTED

1 Suggest three reasons why using a tasting test with a number of people might be a good idea to inform you as to whether any alterations are needed to a recipe that you have trialled. [3 marks]
2 Why might you need to estimate the cost of ingredients for a dish? [2 marks]
3 Identify three ways you might check whether a food is cooked. [3 marks]

Explaining, justifying and presenting ideas about chosen recipes and cooking methods   REVISED

You need to be able to suggest alternative ingredients, processes, cooking methods, calculate portion sizes, and to be able to explain and justify why you have made these changes. Your justification and explanation should include reference to:

- **Any scientific basis for the change:** Using your scientific knowledge about the functions of ingredients, such as coagulation, gelatinisation or dextrinisation, will allow you to explain fully why you have substituted ingredients.
- **Any dietary need for the change:** If you have substituted an ingredient to cater for someone who is lactose intolerant, who is a vegetarian or vegan or who has a dietary illness, you need to explain what the substitute ingredient is and why you have done this.
- **Any taste testing or practice sessions** you have carried out: Refer to the results of these and explain how they have given you relevant information to alter your chosen dish. If you have practiced your dish, and made alterations because of failures in the original recipe, explain what these were and why you have made changes.

- **Any change because of timings or cost:** List what changes you have made and give reasons for each one.
- **Any extra ingredients, seasonings or additions of colour or texture** that you have included: Include information as to why you are making these alterations or adaptations.

Making decisions about which techniques are appropriate to use during preparation and cooking

REVISED

- Your understanding of **nutrition**:
  - ○ Show knowledge and understanding of types, roles and functions of macronutrients and micronutrients. Explain how these nutrients work in our bodies, and what happens if we eat too much or not enough of these.
  - ○ Look at the different energy requirements of individuals, planning balanced diets and calculating energy content and nutritional values of dishes and meals.
  - ○ Look at how heat affects nutrients. You need to show that you can use all this knowledge when deciding how to cook and prepare your ingredients for your practical sessions. If you are trying to conserve nutrients, state which method of cooking does this.
- Your understanding of **food**:
  - ○ Look at the science of food, what happens when it is cooked and how the properties of different nutrients are used to create the dishes we cook. You need to show that you can alter the techniques or type of ingredient to fulfil the brief or recipe.
- Your understanding of different **culinary traditions**:
  - ○ You need to be able to alter a recipe or ingredient to cater for a different religion or culture that has sets of rules for the food that they can eat.
- Your understanding of **food preparation and cooking**:
  - ○ You will have tried out many recipes and techniques when making food. Each time, your evaluation should have included what went well, and worked for you, and what needed improvement. Use this information to select and justify appropriate recipes and dishes.

> **Exam tip**
>
> A question on this section could ask you to give scientific examples of when it might be necessary to alter the ingredients in a recipe. An example of this could be that you would add an extra ingredient to improve the setting or coagulation in a particular dish. In this case you could answer: 'My quiche was not set when I removed it from the oven during my practice cook. I decided that because I was using skimmed milk, which has a high water content, it would be necessary to add an extra egg the next time I cook this dish. Because eggs are a protein food, and the denaturation of the egg during the cooking process means that the egg changes form as the DNA is unravelled and reassembled, the egg coagulates and sets. By adding an extra egg, it will improve the outcome of the quiche and hopefully ensure the filling is set correctly'.

Now test yourself

TESTED

1 Give two examples of situations when it may be necessary to alter the nutritional values of a recipe, and explain how you would do this in each of your chosen examples. [4 marks]
2 Explain why glazing a product improves the appearance. [2 marks]
3 Suggest two ways you could increase the fibre content of a dish. [2 marks]

Success in the examination

The exam paper

REVISED

The final exam is worth **40%** of your total mark.

You will be tested on everything you have learned during your course:
These topics are:
- Food commodities
- Principles of nutrition
- Diet and good health
- The science of food
- Where food comes from
- Cooking and food preparation.

Your answers will be assessed against the assessment objectives (AOs).

	Assessment objective (AO)	Weighting for written exam
AO1	Demonstrate knowledge and understanding of nutrition, food cooking and preparation.	15%
AO2	Apply knowledge and understanding of nutrition, food, cooking and preparation.	15%
AO3*	Plan, prepare, cook and present dishes, combining appropriate techniques.	0%
AO4	Analyse and evaluate different aspects of nutrition, cooking and preparation including food made by yourself and others.	10%
*You will not be assessed against AO3 in the written exam, only AOs 1, 2 and 4.		

Exam tips

When you are allowed to open the exam paper:
- Read the instructions on the front of the paper.
- Use the correct colour pen.
- Read the paper through from start to finish and begin with a question you know the answer to.
- Make sure you read each question carefully (read it twice), so you understand exactly what the question is asking for.
- Underline or highlight key words.
- Look at the number of marks available and see how many answers or points you have to give.
- Have a go at all of the questions. You may pick up extra marks.
- Understand key words in the question, as your answers will have to be structured in a different way, depending on those words.

Key word	How to answer the question
Identify/suggest/give a reason for	Make a list; write a short answer; select words from a diagram or table to complete gaps in a sentence.
Describe	Make a detailed explanation as to how and why something happens.
Explain	Clarify a subject or point by writing down the meaning of it and then showing you understand it by giving reasons.
Analyse	Break an issue down into its separate parts and look at each part in depth, using evidence and explanations to show your understanding.
Evaluate	Make a judgement about how successful or unsuccessful something is and say why it is important. Include evidence for your answer, and come to a final conclusion.
Discuss	Write about all evidence for and against a topic, or point out the advantages and disadvantages of a topic. Use evidence to arrive at a conclusion.

Plan out your answers before you begin writing with a spider diagram to make sure you do not miss any points.

Example of a spider diagram for a question that has asked you to identify and explain which key factors have an impact on food security:

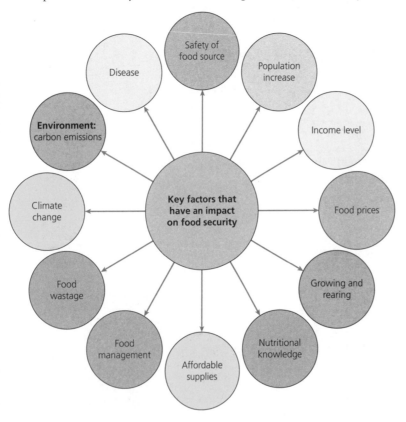

Types of question

The paper will have a mix of questions.

Data-response questions

- These will give you a table of information, a graph or a pie chart and ask you questions on the information shown.
- Refer to the data in your answer and explain how you used it.
- You should also use your own knowledge to answer questions.
- This type of question will usually give one mark for each piece of information you give from the data, and one mark for your own knowledge.

Now test yourself answers and quick quizzes at **www.hoddereducation.co.uk/myrevisionnotes**

For example, you could have a table showing how many teenagers eat fast food every day. The number of marks available is 2. You would answer the part of the question on the data. The next part of the question could ask you why it is not a good idea to eat fast food so often. You would give your own knowledge in this part. You gain one mark for correctly identifying the number of teenagers that eat fast food every day, and one mark for your own knowledge as to why it is not a good idea to eat fast food every day.

Structured questions

These questions give you a piece of information and ask you questions about it.

For example, you could be given a recipe. The questions would ask you to identify ingredients that are not suitable for a vegan and to suggest alternative ingredients, explaining why you chose those alternatives.

- Make sure you use your knowledge and write clearly, giving factual reasons for your reasons and opinions.
- Include as much information as possible.
- You can use the extra sheets at the back of the exam paper, or ask for extra paper if there is not enough room for your answer. If you do this, make sure you label the question with the correct number, and write at the bottom of the original question section 'continued at the back of the paper', or 'continued on extra answer sheet'.

Free-response questions

These ask you to write in more detail about a particular topic, but you will include your own knowledge to provide facts, examples and opinions. You would write your answer in your own way.

These questions will start with one of the words from the key words table on the previous page:

- Describe
- Evaluate
- Explain
- Discuss.
- Analyse

Make sure you understand what the question is asking for. Look at the number of marks available.

- This is the type of question that needs a plan prior to answering.
- Keep referring back to the plan to make sure you do not miss any of the points you have made.
- Use scientific terminology where relevant to increase the marks you gain.
- Always have a go at these questions.

These types of questions would be marked using the different AO bands listed above, so you may have to show AO1, knowledge and understanding of nutrition, and AO2, apply knowledge and understanding of nutrition, to get the full marks for the question.

You can use the extra answer sheets at the back of the exam paper, or ask for extra paper if there is not enough room for your answer. If you do this, make sure you label the question with the correct number, and write at the bottom of the original question section 'continued at the back of the paper', or 'continued on extra answer sheet'.

Don't forget to check all of your answers after you have written them. Use any time you have left after you have finished the paper to read through your answers and add any additional information.

Sample examination questions, model answers and mark schemes

This section includes some sample exam questions, the mark schemes that go with the questions and two sample answers, one of which will get high marks and one of which will get lower marks.

Question One

Look at the following labels listing ingredients for a white loaf of bread and a multi-seeded loaf of bread. Answer the questions beneath.

> **White bread** Fortified Wheat Flour (**Wheat** Flour, Calcium Carbonate, Iron, Niacin, Thiamin), Water, Yeast, **Wheat** Gluten, **Wheat** Fibre (2%), Fermented Wheat Flour, Salt, **Soya** Flour, Spirit Vinegar, Rapeseed Oil, Emulsifier: Mono- and Diacetyl Tartaric Acid Esters of Mono- and Diglycerides of Fatty Acids; Palm Oil, Flour Treatment Agent: Ascorbic Acid.

> **Wholemeal bread** Wholemeal Wheat Flour, Water, Sunflower Seeds (6%), Brown Linseed (4%), Millet (3%), Wheat Gluten, Yeast, Poppy Seed (2%), Sugar, Rapeseed Oil, Salt, Soya Flour, Fermented Wheat Flour, Malted Wheat and Malted Barley Flours, Spirit Vinegar, Rye Flour, Palm Oil, Flour Treatment Agent: Ascorbic Acid.

1 (a) Identify two ingredients in the white sliced bread that are not listed in the ingredients for the wholemeal bread. [2 marks]

(b) Explain why the white bread has fortified wheat flour. [2 marks]

Example

High-mark answer

(b) When wheat is processed to make white flour, the outer layer or bran is removed from the wheat. This means that a lot of the beneficial nutrients are removed, so the resulting white flour will have these nutrients re-added to it after processing, and will be fortified with these ingredients to make sure that people who choose to eat white bread rather than wholemeal bread still get the correct nutrients.

Assessment comment

This answer would gain full marks as the student has explained what happens during the processing of the wheat to make white flour, and has explained what fortified means.

Example

Low-mark answer

(b) Fortified means that extra ingredients have been added to make the food more nutritious.

Assessment comment

In this case the student has not answered the question, which asks for why the white bread has been fortified, not what the word fortified means.

Mark scheme

To gain full marks:

1 (a) Any two from: Fortified wheat flour (wheat flour, calcium carbonate, iron, niacin, thiamin), vegetable oil (rapeseed, palm), soya flour.

Any other answer will not gain a mark. Any other ingredient mentioned from the label will not get a mark.

(b) 2 marks for a full answer mentioning the following terms: processing of wheat, bran, loss of nutrients during processing, reintroduction of lost nutrients

1 mark for knowing extra nutrients are added but with little or no explanation of why or how.

0 marks for an answer that does not mention removal or addition of nutrients.

Question Two

REVISED

You are making a chicken and mushroom pasta bake. The main ingredients in your recipe are:
- raw chicken breast
- mushrooms
- milk
- flour
- butter
- pasta
- breadcrumbs and Parmesan cheese for a crunchy topping.

(a) Explain how you would avoid possible food poisoning when:

(i) shopping and transporting ingredients home [2 marks]

(ii) preparing the ingredients for the pasta bake [2 marks]

(iii) storing the pasta bake and then reheating later that evening. [2 marks]

(b) Suggest one way you might make the pasta bake look attractive. [1 mark]

Example

High-mark answer

2 **(a) (i)** When shopping for the food I would always check the sell-by date before buying chicken. I would place the chicken in a cool bag to transport it home to prevent it warming up and putting it in the danger zone where bacteria can begin to grow.

Assessment comment

(i) Full marks have been awarded for giving two reasons with explanations.

Example

(ii) When preparing the ingredients for the pasta bake, I would use a separate knife and board for the raw chicken to prevent cross-contamination, and would wash my hands before and after preparing the raw chicken to avoid cross-contamination.

Assessment comment

(ii) Full marks have been awarded for giving two reasons with explanations.

Example

(iii) When storing the pasta bake, I would make sure it cooled down completely before covering it and placing it in the fridge away from any raw foods. This means that the dish will not be left in the danger zone, and it cannot be contaminated by raw food in the fridge.

Assessment comment

(iii) Full marks have been awarded for giving two reasons with explanations.

Example

(b) To make the pasta bake look attractive, I would chop parsley finely and sprinkle it on top of the dish prior to serving, to give a colour contrast to the final appearance.

Assessment comment

One mark has been awarded for a suitable suggestion.

Sample examination questions, model answers and mark schemes

Example

Low-mark answer

2 (a) (i) When shopping you need to buy the best chicken and take it home and put it in the fridge.

Assessment comment

This has not gained any marks, as the student has not given a valid answer for buying the chicken, and has not talked about transporting it home.

Example

(ii) When preparing the chicken I need to wash my hands to make sure they are clean.

Assessment comment

This has gained one mark, as only one point has been given.

Example

(iii) If I reheat the pasta bake it must only be reheated once as the bacteria keep growing after the dish has been cooked and left to cool down.

Assessment comment

This has gained two marks, as the student has mentioned reheating once, and the reason why.

Example

(b) To make the dish look attractive you can add sliced tomatoes on top before cooking to give colour.

Assessment comment

This has gained one mark, as it is a suitable suggestion.

Mark scheme

(a) Award 1 mark each for any two correct responses making reference to:
(i)
- packaging intact
- best-before date
- fresh appearance – mushrooms not wilted or rotten
- transporting chicken home in a cool bag or wrapped separately.

(ii)
- personal hygiene
- prevention of cross-contamination
- use of colour-coded knives and chopping boards
- correct washing-up technique.

(iii)
- waiting until the pasta bake has cooled down before putting it in the fridge
- covering with cling film before placing in the fridge
- placing on a high shelf so no raw ingredients can drip or contaminate the cooked food
- reheating to a high temperature for sufficient time to make sure it is piping hot all the way through
- only reheating once.

(b) Award 1 mark for a suitable suggestion to make the dish look attractive. Examples could be:
- add chopped parsley to give colour
- add sliced tomatoes prior to reheating to give a colourful addition
- serve with a salad garnish or green vegetables to add a contrast in colour.

Question Three

3 Everyone requires different amounts of energy.

(a) Describe the factors that will influence how much energy an individual will require. [6 marks]

(b) Explain what happens if there is an imbalance between energy input and energy output. [6 marks]

Example

High-mark answer

3 (a) There are a great many factors that will influence how much energy an individual needs.

Energy needs are calculated by using the equation:

Basic Metabolic Rate (BMR) x Physical Activity Levels (PAL) = Estimated Average Requirement (EAR)

Your BMR is the energy needed to keep your body breathing, making chemicals, keeping your heart beating and other body organs working and to keep your blood pumping and nerves working.

Your PAL is the amount of exercise you do, and varies from less than 1.4, which is a bed-bound hospital patient, to 2.4, which is a professional athlete, such as a footballer.

Your BMR will be different depending on what age you are and whether you are male or female. A young child's BMR will be much less than a full grown adult, due to the size of the person.

PAL levels will increase the more exercise you do. It is possible to increase your PAL by simply increasing the activities you do, for example walking instead of taking the lift. What occupation you do will also change your PAL, someone sitting at a desk or driving a taxi all day will have a lower PAL than a construction worker, such as a brick layer. This means that the bricklayer will require more energy, as he or she will have a higher EAR from the equation above.

If someone has been unwell, or has had an operation, they may need increased energy provision to allow them to heal properly, or regain lost weight.

Someone trying to lose weight will need to reduce their energy intake to allow the body to burn up some of the excess fat in the body as an energy source.

A pregnant woman will need extra energy to make sure the developing baby has the correct energy to grow healthily. If a woman is breast-feeding, she will need to increase her energy intake to make the milk for the baby.

Assessment comment

This student has gained full marks, as they have discussed at least four different points with a full explanation. The calculation for EAR is included, with explanations of all the abbreviated initials.

Example

(b) Energy balance is when energy input is equal to energy output. This will maintain body weight.

If someone consumes too much energy in their diet, they will begin to put on weight, as the extra energy will be stored as fat in their body. This will increase their risk of developing diseases, for example cardiovascular problems such as Coronary Heart Disease and stroke. It may also cause them to have raised cholesterol levels in their blood. Type 2 diabetes is also linked to being overweight. They may eventually become obese, which will increase their risk of developing cancers and bowel problems.

This may also result in a shorter life span.

People who are overweight or obese have reduced energy levels, and are less likely to exercise because of poor mobility and joint problems. They are more likely to have self-esteem problems, and be embarrassed about their weight. They can be depressed, which can lead to comfort eating, which will exacerbate the problem.

If someone consumes too little energy, they will lose weight, as the body will begin to burn up stored fat to provide energy. There is a risk that the person will become unwell, due to insufficient consumption of the key nutrients required to keep the body healthy. These include protein, fat, carbohydrate, and all the vitamins and minerals. Being severely underweight can cause a female to stop having periods.

A person with insufficient energy intake will become tired and lethargic. They may also have an increased risk of infection or illness due to low immunity levels.

Not consuming sufficient energy can stop the production of breast milk in a lactating female.

> **Assessment comment**
>
> This student has gained full marks, as they have given information about both over- and under-consumption of energy, and have used specialist terms throughout the answer.

Example

Low-mark answer

3 (a) How much energy you need changes because:
  - as you get older you get bigger, so need to eat more energy in food
  - men are bigger than women usually so need more energy
  - if you do lots of exercise, like an athlete, you need more energy.

> **Assessment comment**
>
> This student has gained 2 marks, as they have given three points, in a bullet form but with no explanations of each point. No mention of BMR, PAL or EAR.

Example

(b) If you eat too much food you will put on weight and become obese. This means you can get other things like heart disease. People may make fun of you so you get fed up and eat more. This can make you really ill and get something like diabetes. if you don't eat enough food you will get thinner and lose weight. Some people become really skinny and unwell. Your body burns up the fat and you become thinner. if you don't eat enough good food you get ill because the good things are missing in your food.

> **Assessment comment**
>
> This student has gained two marks, one for saying you will put on weight if you eat too much food, and one for saying you lose weight if you don't eat enough food. There is no specialist terminology used, and no full explanations given.

Mark scheme

Mark scheme for parts a) and b)	**High-level response (5–6 marks)** The candidate presents a well-balanced answer linking directly to the question. At least four different points are mentioned with explanations and specialist terms being used. Accurate spelling, punctuation and grammar are used, with structured sentences. **Mid-level response (3–4 marks)** Mainly answered using structured sentences and specialist terms with mostly correct spelling, punctuation and grammar. At least three points are discussed in detail with some justification and reasoning for answers given. **Low-level response (0–2 marks)** The candidate presents the information as bullet points or a list. At least two points are mentioned. Some basic information is offered to support statements. Errors in spelling, punctuation and grammar are present.

Describe the factors that influence an individual's energy requirements	**Factors that can be mentioned include:** *Age* – different needs at different stages of life; can also link to BMR (BMR – the resting rate of energy usage before activity levels are applied; higher BMR has a higher need for energy) *Gender* – males have a higher energy need than females *Occupation* – manual labour or physical activity at work increases needs *PAL levels* (definition of PAL showing knowledge that increasing activity increases PAL) *Knowledge of EAR* (and the equation that calculates EAR) *Exercise* – more exercise generally requires more energy intake *Illness* – energy is needed for growth, repair, fighting infections *Pregnancy* – increased need as providing for a growing foetus *Lactation* – providing milk for the baby and may be more active
Discuss the effects of an energy imbalance	**Factors that can be mentioned include:** *Consuming too much energy can lead to:* increased risk of overweight/obesity/coronary heart disease/heart problems/diabetes/some cancers reduced/low self-esteem lack of energy to do things/sluggish/poor concentration/motivation lack of specific nutrients may lead to the onset of deficiencies such as rickets shorter life span if overweight or obese *Consuming too little energy can lead to:* risk of losing weight plus a reduction in energy levels/increased tiredness/reduced productivity/do not have enough energy to do the things they need to do increased risk of infection not receiving the correct nutrients – risk of deficiencies such as anaemia periods stops risk of stopping or reducing milk production (harm to baby – reduction of nutrients/anti-bodies) if lactating risk of low-weight baby if pregnant

Question Four

REVISED

4 Food labels provide information for consumers:
  (a) Identify four pieces of information that are required by law to be on a food label, and explain why each item is required. [8 marks]
  (b) Many foods contain additives. Give three reasons that additives are in food. [3 marks]
  (c) Discuss whether additives are beneficial or harmful, giving examples and reasons for your arguments for and against. [6 marks]

Example

High-mark answer

4 (a) The government has made it the law to provide certain information on food packaging, to give the consumer information to make a reasoned choice and be safe.

Four pieces of information required by law are firstly the name and address of the company that produced the product. If there is any problem with the quality of the product, any foreign bodies such as metal or a hair are found in the product, or if the product causes illness, the consumer knows who to contact to complain about the product.

There has to be a list of all ingredients, from the heaviest to the lightest in order of weight. This means that the consumer can check what is in the food and see if they are allergic to any products, or if for some reason they do not want to buy it (for example it contained sugar, and they were trying to lose weight).

There has be storage instructions, such as 'keep chilled in the refrigerator until needed'. This is because the food may go off if incorrectly stored, and the consumer could become ill as a result of eating contaminated food. The manufacturer would then be liable, as they had not told the consumer to put the food in the fridge.

There has to be allergy information, and the allergens that must be included are on a government list. The list includes nuts, which can cause a life-threatening condition when the person has a severe allergic reaction called anaphylactic shock. This is where the throat swells and prevents the person breathing.

Assessment comment

The student has gained all 8 marks because they have listed and fully explained four things that have to be on food labels by law.

Example

(b) Additives are put into food for several reasons.

Three of these reasons are; as a preservative. This means that the additive will extend the shelf life of the product, as it helps prevent the growth of food spoilage micro-organisms such as bacteria, mould and the action of enzymes in the food.

Another additive is a flavour enhancer or intensifier. This is added to make the food tastier, or to replace taste that may be lost during the processing of the food product.

A third additive is a colour compound. This is added to make the food look more attractive, for example red jelly, or to replace a colour that is lost during processing. For example, tinned peas would be grey after processing, so the green colour is put back as an additive, because people would not eat grey peas.

Assessment comment

The student has gained 3 marks, as three different additives are listed with reasons for their inclusion in the product.

Example

(c) Additives have both advantages and disadvantages.

One of the advantages, or benefits, is that they give a consumer more choice. If the person wants to buy some fruit to use in three days' time, because they will not have enough time to buy it on the day they need it, the person can buy fruit that will stay fresh for that long, as it will have a preservative in the package in the form of a small sticker that helps keep the fruit fresher for longer. This gives a consumer a wide choice of foods all year round.

Another advantage is that flavour intensifiers can improve the taste of a product, making it more attractive to a consumer. During processing, some flavours of foods can be lost, so additives can restore that flavour to give a much tastier product. This improves or restores the original features of the product. Added flavours can also create a different product, such as different flavoured crisps.

A third advantage is the use of stabilisers and emulsifiers which make foods that do not ordinarily mix together, like oil and water, stay together in an emulsion. This means that mayonnaise will not separate out in a jar, for example. This makes the product more attractive to the consumer. Emulsifiers also make a food taste creamy and smooth, and sometimes extend the life of a product.

Finally, colourings are added to a product to make it colourful and attractive, or to replace colours lost in processing. This makes eating the product more enjoyable to the consumer. This improves or restores the original features of the product.

The disadvantages of using additives are that some can produce allergic reactions in people, such as a rash.

Another disadvantage is that over-use of additives can disguise a low-quality product and make it appear better than it is. 'Fruit flavour' for example does not have to contain any fruit at all.

Finally, some colourings used as additives have been shown to produce hyperactivity in children. This is obviously a problem.

Sample examination questions, model answers and mark schemes

Assessment comment

The student has gained all 6 marks here as they have presented a set of advantages and disadvantages, with excellent reasoning of each point.

Example

Low-mark answer

4 (a) pieces of information by law are:
- allergy information if you are allergic to anything
- list of ingredients so people know what's in the food
- how to cook it.

Assessment comment

This student has gained 3 marks as they have correctly listed three pieces of information, but with no explanations given.

Example

(b) Additives are used to colour food to make it look better, make it taste better and to make it last longer.

Assessment comment

This student has gained 1 mark for a simple list with no extra explanations.

Example

(c) Advantages of additives:
- keeps food fresh for longer so it can be stored for a while
- makes it taste better and look nice so it is better for the person eating it
- can be used to mix foods like salad dressing.
Disadvantages:
- Can make children hyper.

Assessment comment

This student has gained 2 marks for a simple list with no explanations.

Mark scheme

(a) Information required by law	Answers to include: ● ingredients: listed in order of weight, high to low ● best-before/use-by/sell-by dates ● name of the product ● cooking instructions if necessary ● storage instructions ● allergy information: there is a government list of all allergens that must be listed ● manufacturer's name and address ● country of origin/where made ● weight/quantity information ● lot number of the food so it can be traced back to origin if imported from outside the EU ● a warning if GM ingredients have been used ● a warning if the product has been irradiated ● the words 'packaged in a modified atmosphere' if a specific gas is used in the packaging ● any necessary warnings: a government list of chemicals and ingredients need specific warning words attached.

Marks awarded	**High-level response (7–8 marks)** Four points mentioned with full explanation for each point using structured sentences – no bullet points. Specific terms used appropriately with a well-presented and balanced answer. **Mid-level response (3–6 marks)** At least three points mentioned with reasonable explanations for each point using mainly structured sentences. Some specialist terms used. **Low-level response (0–2 marks)** At least two answers included, possibly as bullet points with some explanation. No specialist terms used.
(b) Reasons additives are used in food	One mark for each correct answer from: ● as a preservative to extend the shelf life of the product ● as a flavour enhancer/intensifier to improve the taste of a food or replace flavours lost during manufacture ● stabilisers and emulsifiers to help foods mix together and prevent them separating out when stored, to give foods a creamy and smooth texture or to extend shelf life ● colourings make foods look attractive, to boost the colour of a product or to replace colour lost during manufacture. Award one mark for a simple list with no explanations included.
(c) Discussion on the benefits and harm of additives	Answers to include: Benefits: ● give consumers a wide choice ● keep foods safe for longer ● improve colour/flavour of a food ● restore original colour/flavour/nutrient levels ● produce a particular flavour, e.g. in crisps ● to produce a particular effect such as a creamy mouthfeel. Possible harmful effects: ● additives can be used to disguise low-quality ingredients ● people can be allergic to some additives, causing skin rashes, for example ● some additives can cause hyperactivity.
Marks awarded	**High-level response (5–6 marks)** The candidate presents a well-balanced answer linking directly to the question. Points for both the benefits and harmful consequences of additives are included. Good reasoning for answers and ideas. A whole range of specialist terms are used with precision. The candidate can demonstrate the accurate use of spelling, punctuation and grammar. The response will be well presented and in structured sentences throughout. **Mid-level response (3–6 marks)** The candidate will use mostly structured sentences, and both benefits and harmful consequences are discussed. Specialist terms will be used accurately. The candidate will offer some justification and reasoning for answers given. Some spelling, punctuation and grammar errors. **Low-level response (0–2 marks)** At least two benefits and two harmful consequences are included, possibly as bullet points with some explanation. No specialist terms used.